D0842601

To readers with the curiosity to understand
our world and the ambition to improve it

www.mascotbooks.com

The Human Condition

©2019 Gregory A. Loew. All Rights Reserved. No part of this publication may be reproduced, stored in a retrieval system or transmitted in any form by any means electronic, mechanical, or photocopying, recording or otherwise without the permission of the author.

Author photo by Amanda Austin.

The views and opinions expressed in this book are solely those of the author. These views and opinions do not necessarily represent those of the publisher or staff.

For more information, please contact:
Mascot Books
620 Herndon Parkway, Suite 320
Herndon, VA 20170
info@mascotbooks.com

Library of Congress Control Number: 2018909139

CPSIA Code: PRFRE0419A
ISBN-13: 978-1-64307-095-7

Printed in Canada

THE HUMAN CONDITION

REALITY, SCIENCE AND HISTORY

Gregory A. Loew

CONTENTS

INTRODUCTION

This book is an essay on the human condition: our amazing accomplishments, some of our worst failures, and what we need to do to achieve a better future for our species. To address these issues, I begin by taking stock of where we are now and explore how evolution has uniquely positioned our species to access the mystery we call "reality." I look at this process in detail and how our curiosity and ingenuity have led to our spectacular scientific discoveries, technological inventions, and artistic creations. Unfortunately, technology by itself is blind: as it solves certain problems, it creates others. I then show that history is chaotic and that we will need some drastic changes to make our world more ethical and peaceful. *Homo* is indeed smart, but not yet *sapiens* (wise) enough.

Today, we humans share our wonderful planet with the plants and the animal kingdom from which we evolved and on which we depend for survival. We are capable of love, empathy, sacrifice, and cooperation. We can plan for the future. We create literature and art, compose and play beautiful music, construct the General Theory of Relativity, find cures for terrible diseases, and land on the moon. At the same time, we sit at the top of the food chain as the biggest predators on earth. As a species we can hate; discriminate against various groups, religions, and races; mistreat women; abuse drugs to escape reality; exploit and inflict suffering on animals; create chaos and anxiety through terrorism; displace and kill millions of people; and threaten each other with nuclear, chemical, and biological weapons. For a variety of reasons, nations get into interminable civil and international wars which the U.N. is incapable of controlling. To make things worse, worldwide cyber warfare and computer hacking operations that infiltrate social media networks are now used to undermine our democratic institutions and steal commercial secrets. The world population, currently hard to support with 7.5 billion people, is likely to increase to 10 billion or more. We are all thermodynamic machines and use increasing amounts of energy and raw materials per capita to survive and satisfy our way of life. The Greeks invented democracy and built the Parthenon, but spent much of their time at war with the Persians and amongst themselves. In the end, Athens was destroyed. Today, extreme ideologues destroy places like Palmyra and Aleppo. Driven by ignorance, fear, and economic insecurity, even democracies elect incompetent demagogues that mislead them. There is a significant risk that in the future we will irreversibly harm our global environment for lack of sound policies, perhaps use nuclear weapons by intent or by mistake, and make life on earth unlivable.

I must acknowledge that as I was completing this book, I came across two remarkable books titled "Sapiens: A Brief History of Humankind[1]" and "Homo Deus: A Brief History of Tomorrow,[2]" both by Yuval N. Harari. As can happen when writers treat similar subjects, I seem to share some ideas about our past and concerns about our future with him and I will point them out in my book. But as readers familiar with Harari's work will soon find out, my perspective is not quite the same as his and it has led me

1. See Yuval Noah Harari, "Sapiens: A Brief History of Humankind," HarperCollins Publishers, New York, 2015.

2. See Yuval Noah Harari, "Homo Deus: A Brief History of Tomorrow," HarperCollins Publishers, New York, 2017.

to investigate the status of our human condition from somewhat different angles.

This book comes in five separate parts and a conclusion:

- In Part 1, I take a multidisciplinary journey involving philosophy, evolution, psychology, language, and the sciences to explore the nature of reality as we experience it.
- In Part 2, I show that reality cannot be described as a single concept and I categorize four different ways in which we access it, think of it, view it, and study it: *representative reality, mental subjective reality, objective reality, and scientific reality.*
- In Part 3, I summarize with various selected examples the major progress we have made in mathematics and the various natural sciences, and how this knowledge contributes to our current understanding of the universe. I also discuss the progress that has been made in the social sciences, as well as how they have been unable so far to come up with solutions to our problematic individual and social behaviors.
- In Part 4, I look at some of the highlights of history: our checkered past, our repeated mistakes, and our glorification of war. The course of history, which has been recorded for three thousand years, is irreversible and its forward motion keeps changing in large part because of scientific and technological developments. But unlike science and technology, history cannot be repeated and tested by experiments. I also show how history is punctuated and affected by the enactment of new laws, and that these laws often lag behind the needs of society.

- In Part 5, I look at the future of the human condition in the short and long run and how we might influence and improve its prospects. I examine the conditions under which human nature might or might not change, and the attendant risks of tinkering with our genes and our mental faculties. I try to guess where science and technology might take us in the future. I then review the universal problems faced by our education systems and the changes we could make to remedy some of our social woes. In the end, I propose admittedly utopian ideas on how and where we might make improvements to our economic, political, and international problems.
- Finally, in my conclusion in Part 6, I discuss my position on humanism and happiness.

PART 1:

A JOURNEY THROUGH PHILOSOPHY, EVOLUTION, PSYCHOLOGY, LANGUAGE, AND SCIENCE

The Philosophers and Reality

Unlike renowned physicist Stephen Hawking (1942-2018), who asserted that "philosophy is dead[3]," I believe that philosophy as a discipline is still necessary to enlighten and guide us. This is true even though in the last three hundred years or so, science has displaced philosophy from many fields of inquiry. Philosophy still provides us a framework within which to view all knowledge we acquire (the domain of epistemology) and all modes of existence of which we are aware (the domain of ontology).

Throughout history, philosophers from Democritus (~460-370 BCE) and Plato (~428-347 BCE) to the present have held a wide spectrum of world views on reality, ranging from pure idealism to pure materialism. Pure idealists (sometimes called "subjective idealists"), such as Bishop George Berkeley (1685-1753), believe that only thoughts "really exist" (You, the reader, are not really there for me unless I think of and am aware of you!). Pure materialists, such as Thomas Hobbes (1588-1679), believe that only matter and energy "exist" and that our minds and states of consciousness are integral parts of the material world, even though we don't yet understand (in our consciousness) how this works. Between these two extremes, many intermediate philosophical positions ("-isms") such as empiricism, rationalism, direct and indirect realism, positivism, and so on, developed over time.

These positions evolved as successive philosophers reviewed and modified the ideas of previous individuals or groups. Many philosophical ideas were derived from introspection, intuition, belief, and the occasional need to accommodate or oppose contemporary religious dogma and/or scientific discoveries. The positions also contained different epistemological views. For example, em-

3. See Stephen Hawking and Leonard Mlodinow, "The Grand Design," Bantam Books, 2010, page 5.

piricist David Hume (1711-1776) believed that all knowledge is obtained from our sensations, and that our minds are *tabula rasa* ("blank slates" devoid of any knowledge) unless exposed to the messages from our senses. He thought that cause-and-effect relationships could only be inferred by observation, not by reason.

Fig.1 Immanuel Kant (1724-1804) who thought we only experience phenomena through our senses but never the "noumena" or things in themselves.

On the other hand, idealists and rationalists emphasized the inherent contribution of our minds to the creation and formulation of knowledge. Towering over his predecessors, Immanuel Kant (1724-1804) combined these ideas and stated that our minds lend structure to both the necessary *a priori* logic of mathematics and the experiences derived from our sensations. Originally interested in physics and astronomy, he formulated the theory that we never really get to know the intrinsic or *noumenal* elements of the world (the things in themselves) but, rather, experience its *phenomenal* elements through our senses, which we can analyze. Would Kant think differently today if we told him what we know about the atoms, molecules and fundamental particles that make up our *noumena*[4]?

4. Arthur Schopenhauer partially disagreed with Kant and said that when it comes to our bodies, which we "inhabit," we do actually know the "thing in itself."

No, because as we will see later, our knowledge of these constituents is still derived from *phenomena*.

In his epistemological analysis, Kant described a hierarchy by which we acquire, formulate, and assimilate knowledge, which can be depicted as follows: raw sensations ("Ouch, it's hot!"), selective perceptions ("There is a hot stove here"), conception (the abstract idea of heat as a property), and ultimately intention ("I won't burn myself on a stove next time!"). Regarding "space and time," Kant thought that they are not independent entities, but rather the mental frameworks that allow our brains to organize our perceptions. I will come back to "space and time" later.

During the 20th century, the explosion of science triggered the development of a more specialized branch of philosophy dedicated to the study of the foundations of science. Well-known authors like Hans Reichenbach (1891-1953), Rudolf Carnap (1891-1970), Karl Popper (1912-1994), Thomas Kuhn (1922-1996), and others left their imprint[5] in this field. One of the basic questions they asked themselves was, "what makes a theory scientific?" Some believed that the necessary criterion was that a theory had to be verifiable through observation and experiment. But for how long would observations and experiments have to be performed? Popper came up with a sharper line of demarcation, the criterion of "falsifiability:" *i.e.*, to be scientific, the theory must contain some explicit predictions that can be shown to be wrong[6]. Such a demarcation works well for simple statements like "all swans are white," which is "falsified" as soon as a black swan is found to exist. On the other hand, it is not clear that scientific theories can always be formulated in this manner, nor that

5. A comprehensive review of 20th century philosophy of science can be found in Peter Godfrey-Smith, "Theory and Reality, an introduction to the philosophy of science," The University Press of Chicago, 2003.

6. Popper also suggested the idea that there are three worlds of knowledge, the physical world, the subjective world of our perceptions, and the objective abstract world of our creations.

they evolve along such a pattern. Kuhn stated that theories develop within a certain framework of assumptions or "paradigm" until it fails to explain new data. At this point a "paradigm shift" takes place through a major transition to a completely new framework. The reader will find later in this essay that I believe that Kuhn is generally correct, although the "shifts" may not always be obvious.

I now want to discuss how evolution has shaped our perceptions and our brains, and developed our capability of analyzing and understanding the world.

Evolution and Reality

Fig.2 Charles Darwin (1809-1882), one of the two fathers of the theory of evolution, and author of "The Origin of the Species by Means of Natural Selection" (1859).

The Theory of Biological Evolution as we understand it today can indeed give us insights into the above philosophical discussions and into a number of other issues to be discussed in the pages to follow. When it emerged, this theory marked a definite paradigm shift with respect to all the preceding Aristotelian, "Intelligent Design," and "Creationist" explanations of the multiplicity of living species on earth.

Fig.3 Alfred Russel Wallace (1823-1913), the other father of the theory of evolution.

By 1858, Charles Darwin (1809-1882) had been developing his theory of "natural selection" for over twenty years when another scientist, Alfred Russel Wallace (1823-1913) who, while in Indonesia, sent Darwin his own paper with essentially the same theory. As a result of friendly interventions of third-party scientists, it was agreed that their separate reports on the subject would be jointly presented and read, in absentia of both, at a meeting of the Linnean Society of London on July 1st, 1858. Wallace was still in Indonesia, and Darwin was attending his son's funeral. After this, Darwin published his magnum opus, *On the Origin of Species by Means of Natural Selection*, in November 1859. What his theory proposed was that all species on earth descended from each other and a common ancestor, that small changes took place within individual species over successive generations, and that large-scale changes over longer periods of time would lead to new species. Darwin did not know where the variations of traits within a species came from since he was not aware of the work first published in 1866 by Gregor Mendel (1822-1884) on the laws of inheritance, the existence and behavior of genes, nor about the effects of mutations due to radiation or chemical effects on these genes. However, he concluded correctly that members of a species with a trait better suited to their physical environment would have more offspring and hence become more numerous through "natural selection."

One such trait, common to all animals, is that to survive they must have acquired some form of sensory and motor apparatus of increasing complexity over time to interact with their environmental "reality." This is true for the worm navigating in a burrow using its sense of smell, the insect eating a leaf, the frog catching a fly with its tongue, the bird building a nest, the lion killing an antelope, an orangutan extracting larvae from a tree with a twig, and early humans hunting bison and covering themselves with their hides. Members of all these species sense and perceive the world and cope with it at a given *scale*, commensurate with their bodies' capabilities and their environment.

Fig.4 Gregor Mendel (1822-1884), Austrian Augustinian friar from Moravia, considered as the founder of modern genetics through his discoveries of the laws of heredity of the traits of garden peas.

Note that most mammals and humans, in addition to their five senses of sight, hearing, taste, smell and touch, which serve as their windows to the outside world, also have a variety of other senses: pain, hunger, thirst, pleasure, sexual impulses, balance, temperature that their bodies communicate to their brains, and senses of equilibrium, kinesthetic motion and direction that their brains send to their bodies. As humans, we needed to perceive animals and plants in the jungle or the savanna to survive, but we did not need eyes

Fig.5 Killer whales (orcas) are very friendly to humans but are ferocious hunters. So-called "transient" orcas eat only mammals like sperm whales, dolphins, seals and penguins, while "resident" orcas eat only fish like salmon and herring.

that could see atoms or the moons of Jupiter (although it would be nice if our eyes could allow us to see viruses!).

How animals and humans evolved and acquired their innumerable faculties, including the awareness of their bodies and the external world, and their sense of "self" and consciousness[7] will no doubt continue to be a source of fascination and research for many years. As humans, we sometimes act as if there were an absolute discontinuity between our species and those of all other animals. This misconception is admirably dispelled by Carl Safina in his recent book *Beyond Words: How Animals Think and Feel*[8].

7. See for example John Searle's "The Mystery of Consciousness Continues" in his critique of neurologist Antonio Damasio's "Self Comes to Mind, Constructing the Conscious Brain," Pantheon, 2010, in the New York Review of Books, June 9, 2011. Searle believes that all our mental phenomena are caused by lower level neuronal processes in the brain and are themselves manifested as higher level features of consciousness. Searle does not believe in the traditional mind-body dualism.

8. See Carl Safina's "Beyond Words: What Animals Think and Feel," Henry Holt & Co., 2015.

In describing the intimate lives and behaviors of elephants, wolves, apes, dogs, birds, and the amazingly clever and sociable dolphins and killer whales (orcas), he shows how close we really are to them genetically and psychologically. To quote him, as humans we should show some perspective and humility: "Twenty-five million years before today, dolphins were firmly in possession of our solar system's brightest brain...When dolphins were the planet's brain leaders, the world didn't have any political, religious, ethnic, or environmental problems..."

And yet, modesty aside, we humans have now become unique. Around five-and-a-half million years ago, we became separated as a species from the chimpanzees with which we share a large percentage of our genes[9]. From that time on it is now thought that we first evolved very slowly in Africa from *Australopithecine* hominids like Ardi[10] and Lucy[11].

About two and a half million years ago, as genus *Homo Habilis,* we started to make stone tools and hunting weapons. This time defines the beginning of the Stone Age, during which we became bipedal and our brains grew enormously in size. Between 300,000 and

Fig.6 Contemporary chimpanzee in a zoo.

9. The exact percentage, which is sometimes quoted between 94 and 98.5%, is highly controversial because it depends on how the detailed similarities and differences between corresponding genes in the two species are taken into account.

10. See White, T.D. et al, 2009, Ardipithecus ramidus and the paleobiology of early hominids, Science, 326, 75-76.

11. See Johanson, Donald and Maitland, Edey, "Lucy: The beginnings of Humankind," Simon and Schuster, New York, 1981.

200,000 years ago we evolved to what is now called *Homo sapiens*, the supposedly "Wise One," and we then gradually spread all over the world.

Until about 10,000 BCE, we lived in small bands of hunter-gatherers, vulnerable to the elements and predators. We survived as omnivores, eating vegetable products, fruits, and meat. We may have exterminated some species of animals in the process but remained in fairly stable equilibrium with our environment. But then, in what is called the Neolithic Age, we invented agriculture and the domestication of animals, and we became more sedentary. Yuval Harari (Ref.1) believes that this was one of our first huge mistakes although it seems to me that we also became slightly less vulnerable to the elements. For better or for worse we began to exploit our environment more aggressively. The Bronze and Iron Ages subsequently brought about new technology and a population explosion. This was probably the key turning point in our destiny. We developed language, culture, writing, art, laws, religion, and science, and in the process sharpened our view of reality. Unfortunately, along with these developments came autocratic leaders, slavery and human exploitation, wars, conquests, nations, and empires. As I said earlier, *Homo sapiens* turned out to be very smart and resourceful but **not wise enough**.

Our considerable evolutionary growth in brain size, memory, and number of neurons as compared to our ape ancestors gave us enormous advantages in many aspects of life and behavior. Our perceptive systems became more integrated with our brains. With extra memory, we developed the ability to mentally conjure up improved images of the external world (perhaps to protect ourselves in the dark of night when the tigers and hyenas came to get us (sic)) and to begin to create words and formulate rudimentary concepts. We eventually became able to communicate with ourselves and others through language.

Perceptions and Reality

To understand how our sense of "reality" seems to depend on our perceptions, let us begin with an example of our perception of colors: the red in roses, the blue of the sky, and the green of leaves. Painters live by colors and combine them into patterns to make beautiful pictures. You feel the colors are real (unless you are colorblind) and that they exist as such, independently of us. Yet, science tells us otherwise: they are just different ways in which our retinas with their cones and rods respond to electromagnetic waves within a very narrow range of wavelengths. All electromagnetic waves, visible light included, are composed of quanta of energy called "photons" that travel in vacuum at the same speed (c) and are characterized by a wavelength (λ), the distance between two adjacent crests, as shown below:

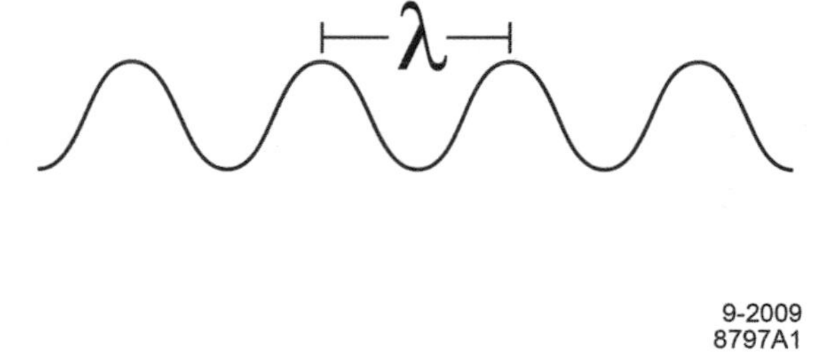

Fig.7 Sinusoidal wave of wavelength lambda.

The human eye is capable of detecting photons of light with wavelengths (λ) ranging from about 0.375 microns (violet) to 0.75 microns (red), where a micron is a millionth of a meter. This information is first processed in a distinguishable way in the retina and then by a very complex mechanism in our brains. Some believe that evolution favored this mechanism in part to help species recognize different foods. At any rate, if it weren't for our ability to distinguish colors (wavelengths), there might have been black and white drawings but most painters would not exist as such. Red, blue, and green as colors have no reality outside of our brains, and yet we know from physics that the corresponding photons are selectively reflected by paintings made with "real" oils representing "real" people and objects.

Fig. 8 "La Grenouillère" by Pierre-Auguste Renoir (1841-1919) representing "real" people and objects.

So, what does this mean? We know that physicists can measure wavelengths. Biologists can measure the cones and rods as the light impinges on our retinas. Neurophysiologists try to follow the complicated neural pathways that transmit the signals from the cones and rods to our central nervous system. They can also do the same experiments with other species and study how their perceptions of colors differ from ours and why. For example, bees don't perceive red.

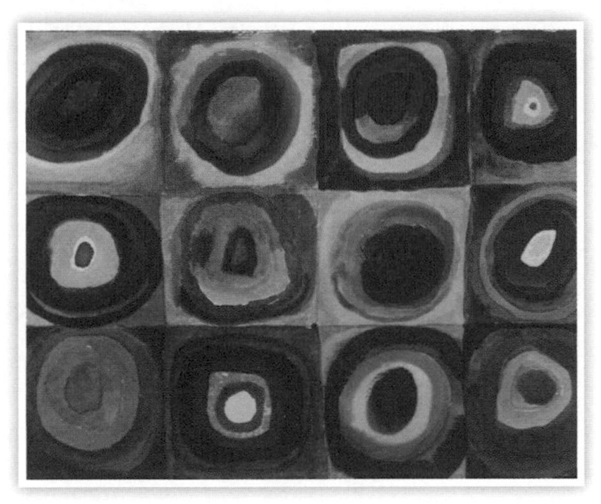

Fig.9 "Circle paintings" by Wassily Kandinsky (1866-1944) representing abstract shapes.

Up to this point, all these processes were measurable. When engineers created digital cameras, they did within a few years something similar to what nature did in millions of years. However, they didn't completely recreate the natural phenomenon to the end! The "finale" is what is not understood yet: how our central nervous system then creates our states of awareness of red, blue and green. Psychologists and philosophers call this mysterious process "qualia," the fact that physical effects become entirely subjective experiences. The effects of which we become conscious "represent" or "correspond to" the wavelengths of the light reflected by the original object, but the redness of the rose is manufactured in our individual heads and we can't even be sure that it is the same for each and every one of us. Note also that the whole process of visual perception takes a fraction of a millisecond so that our awareness of the color of an object happens after we have sensed the object, and is actually a record of a past event.

But colors are just one of a myriad of perceived properties. What about smells, tastes, and touch? What about sounds which we can perceive only in the range between 20 and 20,000 Hertz or cycles per second? What about those beautiful Mozart pieces? We know he composed them in his head and subsequently filled "real" sheets with thousands of notes. And we know that real musicians masterfully move their bodies and instruments to launch into the air the sounds corresponding to these notes. But does music really exist "out there," independently of us? Or is it just various combinations in amplitude and time of air density modulations that hit our ears that only we humans decode as "pleasing" patterns? Indeed, it looks again like evolution has done something very special to us. Studies reported by Aniruddh D. Patel[12], a neuroscientist of music, indicate that almost no other animals except cockatoo parrots respond to musical rhythms.

Fig.10 Wolfgang Amadeus Mozart (1756-1791).

Once we begin to examine these experiences, we must ask ourselves about the nature of all of our perceptions. Not just what we see, touch, hear, smell and taste, but how we feel pleasure and pain, and so on. Everything that is transmitted by our ner-

12. See New York Times, Science Times section, June 1, 2010, page 2.

vous systems is coded in our brains. And with that we can already conclude that pure empiricism, which assumes that our brain is a "tabula rasa," cannot be correct. Indeed, the abilities to discern colors and many other properties of the outside world, as well as the properties and behaviors of our bodies, even if they evolve as we grow up, are part of the "equipment" we receive at birth.

Language and Reality

I owe several of the ideas expressed in this section to four essays by Robert Berwick and Noam Chomsky published in a recent book[13]. Admittedly, I am rephrasing some of their statements and assertions as I understand them. Their central thesis is that the language "faculty" with which we are endowed is unique to the human species. It is a system, internal to individuals, consisting of a potentially "infinite array of hierarchically structured expressions" that interfaces with thought on one side and sound (or gestures) on the other. Hence, this faculty can manifest itself in two separate ways:

(1) In the silent (mental) verbalization of thought which requires a memory where words are stored and some kind of central processor that can assemble these words into sentences, and

(2) In the process of communication with others which requires a motor/sensory or output/input system. For this communication to take place, the central processor must drive the output/input system consisting of vocal cords and the tongue for the production of speech, and connect to ears for its reception. Alternately, the central processor can derive Sign Language through gestures as output and eyes as input.

How and when did we acquire these systems? There are many conjectures but no fully definitive answers. One reason is that the

13. See Robert C. Berwick and Noam Chomsky, "Why Only Us: language and evolution," Cambridge, MA, The MIT Press, 2016.

paleontological and anthropological record is scant. Another is that when we try to study the evolution of a phenotype like the "language organ," we need to have a good knowledge of its properties to guess how they have been selected. This is not the case for the part of the brain which generates language. We don't even know how groups of neurons store information. A third reason is that the genetics involved in the evolution of language are incredibly complicated.

Darwinian natural selection would indicate a very gradual process through a chain of endless "micro-mutations" over at least two million years. However, Berwick and Chomsky believe that the human language "faculty" probably evolved much more rapidly and discontinuously, between 200,000 and the time around 60,000 years ago when a large wave of *Homo sapiens* left Africa. If this is true, other species or sub-species like the Neanderthals and Denisovans whose ancestors left Africa at least 500,000 years ago are unlikely to have had the language "faculty." Berwick and Chomsky believe that what drove us initially to acquire language was not "communication with others" but the selective advantage of an "inner mental tool" that gave expression to our thoughts, intentions, inferences, and plans. I would rather believe that both processes were interwoven in time. Like other animals, humans for a long period of time had used primitive grunts and sounds to communicate with other members of their species. I would guess that these discrete "early words" (*lexical items*), contrary to what Berwick and Chomsky assert (pages 85-86), were indeed *references*[14] to their needs, actions and perceptions of "external" people or objects (not the people or objects themselves). They could be stored in their memories or learned but they had to come from somewhere. It took a few crucial mutations in their much larger brains to modify the processor and activate the language "tool."

14. Philosophers agonize over the word "reference" because in itself it cannot have meaning. The thought behind it is what carries meaning.

For it to work, certain anatomical fiber connections or loops had to be completed between the memory and the processor in the cortex.

As a result of natural selection, for the last 60,000 years the processor has contained an **innate** program which can "merge" (assemble) words into sentences according to the generic rules of something like a Universal Grammar (UG). The sentences are hierarchical (*i.e.*, they contain subjects and predicates), and they can be arbitrarily short or long. Apparently the above anatomical connections in the brain are incomplete in young children, a fact that explains why they begin by remembering words but cannot form syntactic expressions. After the fiber connections are completed, children learn how to speak and understand at least one of the 7,000 existing languages that they are taught. The language is "grafted" onto the UG template. Note that there are two processes at work here: evolution (nature) that endowed us with the language faculty, and development (nurture, culture, etc.) that generated any one of these many different languages through geographical dispersion. Interestingly, it should be added that a sentence may be "syntactically" correct but devoid of any semantic meaning.

How does language work for us today? To answer this question, we must consult the biolinguists, the linguists, the logicians and the semanticists. We know intuitively that we think, that is, we process information and perform logical operations in our heads much faster than when we try to express them with words. Speaking takes some effort and time, whether we talk to ourselves silently (which we do during much of our waking hours) or if we express ourselves vocally or through sign language. Note that in addition to learning how to speak one or several languages, we humans also have to learn how to read, write, and count. Clearly, the appearance of alphabets and numbering systems was a huge transformational step that expanded our view of reality and the external world. It enhanced our chances of survival and eventually led to our ability to handle and convey abstract thoughts, make laws, and develop quantitative models of this world. But note also that while the language faculty enables so much function, its innateness perhaps constrains us in certain ways: we may indeed be partially stuck in our own "language straightjackets."

Explanations, Logic, Causation, and Semantics

Learning to explain a process, scientific or other, with language requires several years of training. One part of this training, aside from learning grammar and syntax in a particular language, is to learn the rules of logic, a branch of philosophy dating back to the Greeks who introduced us to syllogisms. Syllogisms start with two premises that by *deductive* reasoning based on two-valued logic ("something is either true or false") lead to an inevitable **empirical** conclusion. Example: "All old Greek men are bald. Aristotle is an old Greek man. Therefore Aristotle is bald." In contrast, *inductive* reasoning is based on the intuitive observation that a statement might be true, but one can't be entirely certain.

A further part of the training is that an explanatory statement needs to seem rational or causal. What makes an explanation seem "causal," *i.e.*, showing that something is the result of something else? Hume and even modern philosophers of science struggle with the foundations of "causation"[15] and face some difficulties in coming up with a crisp definition. In practice, most of our inquiries and many of our conversations start with the question "why," followed by the answer "because" and a more elaborate explanation of "how." Children become aware of a *sequential*[16] relationship between cause

15. See Ref. 5 above, Chapter 13.

16. Note, however, that we can all be fooled by the fact that two phenomena can be sequential without being causally related.

and effect, practically from birth. Our everyday life and survival would be impossible if we couldn't believe in causality. No branch of engineering could exist without it. Without causality, we could never have gotten to the moon and back.

The semanticists, dealing with meaning, make us aware of the difficulties and traps arising from the imprecise, redundant, ambiguous, multi-level meanings of words. Here again, we are faced with the problem of "*references*" raised earlier. Words are not invented through a rational process that creates a relationship between a sound and a thought or a unique element of "mental reality." They creep into language through usage, need, cultural pressures, and changes in concepts. In many cases, several different words have the same meaning. In other cases, one word has several meanings. The best proof that words are created in unpredictable ways is that a word that exists in one language does not necessarily have a one-to-one translation into another language. For example, the English word "mind" does not have a perfect equivalent in French, except for the word "esprit" which is only an approximate translation. On the other hand, the English verb "to know" has two translations in French: "connaître" and "savoir," a distinction which is very important. "Connaître" is used to refer to a person, a place, an object or a feeling you know. "Savoir" refers to a fact or an idea you know. It is remarkable that this distinction which exists in Latin languages and in German ("Kennen" and "Wissen") didn't make it into English.

One of the founders of semantics was Alfred Korzybski (1879-1950). Korzybski is remembered for seminal contributions to this field. He distrusted Aristotelian two-valued logic, and particularly to the point made here, Korzybski remains famous for his warning to his colleagues and students not to confuse "the map with the territory." In this he made a profound statement about the distinction between the map (in our minds) and the territory (out there in the world). I will come back to this essential point later.

Science, Images, and Models

As evolution has determined what we perceive and how we use language, we can now examine the nature of science. Science progresses through three intertwined processes: observations of phenomena, the formulation of theoretical models to explain them, and the testing of these models through experiments. As we will se in Part 3, theories and models abound in all fields of science, including medicine. Science expects that when a theory or a model is formulated, it must be successful in explaining or at least predicting an observation or the result of an experiment. It must be verifiable, or in certain cases, as Karl Popper demanded, it must be "falsifiable," *i.e.*, one must be able to show that it is wrong. When we try to formulate models of observed phenomena, we commonly base them on our imagination, *i.e.*, the creation of "images and patterns we perceive from the world out there." From initial assumptions or laws we draw consequences, typically with words or mathematical derivations and their surrogates, computer programs. The miracle of the "language" of mathematics is that it is entirely based on *a priori* logic (as noted by Kant) and, per se, has no semantic content until we assign specific physical meaning to its functions and variables. However, it can then "crank out" results like a machine, an amazing process that is further described below.

PART 2:

THE FOUR FORMS OF REALITY

What we have learned from our fact-finding journey through philosophy, evolution, psychology, language and science is that "reality," an abstract word, reveals itself to us in various forms and stages that we will now try to identify.

The First Stage: Representative Reality

The first stage at which we receive information from the world starts in the womb and continues throughout our lives. It comes from our instinctive drives and through our various senses via the perceptive processes described earlier. Our so-called "percepts" of people, objects, occurrences and events do not a priori require memory to be created but they may get committed to memory, and if these "percepts" reoccur because of repeated exposure, they will create some resonance in our minds and the feeling of recognition. Psychologists and philosophers have had endless discussions about whether these perceptive processes are direct or indirect. An excellent review of this controversy is given by Steven Lehar with respect to vision[17]. He argues convincingly that the hypothesis of **direct realism** is physically and neurologically untenable and that our brains actually provide us with an internal **analogical representation** (as opposed to digital) of three-dimensional space and colors of the external world. Referring to the two pictures here (Figures 11 a and b), according to the direct perception on the left, our minds would have to be simple mirrors, which they obviously are not.

On the other hand, according to the process pictured on the right, **indirect realism** does not mean that there is a "homunculus"

17. See Steven Lehar's "The function of conscious experience: an analogical paradigm of perception and behavior," submitted to Consciousness and Cognition (2000), also available at http://cns-alumni.bu.edu/~slehar/webstuff/consc1/submitted.html

Fig. 11a Steven Lehar's graphic representation of "direct reality" where the individual perceives the house and landscape like a mirror.

Fig. 11b Steven Lehar's graphic of "indirect reality" where the brain of the individual creates an internal analogical conscious representation of the perceived house and landscape.

with eyes inside our heads that "sees" this representation, but rather, that the content of a conscious perception **is this representation.**

Following this reasoning, I choose to call the information received through these processes *representative reality*. To recapitulate, we do not perceive light wavelengths when we look at objects; they are *represented* by geometric structures and colors in our minds. This is not only true for objects like the house with the mountain, the sun and the moon shown above, but also for entire events taking place under our eyes. Similar mechanisms take place with all our other sensory perceptions, both external and internal to our bodies.

I want to come back briefly to the perceptions of space and time introduced earlier with reference to Kant. With respect to the perception of space, it turns out that evolution conserves useful traits over very distant species: for example, binocular vision that combines two-dimensional images from each of two retinas into a three-dimensional mental matrix that enables animals (including humans) to judge distance and move around has been handed down over millions of years. A related mechanism has been discovered recently by 2014 Nobel Prize winners May-Britt and Edvard Moser of the Norwegian University of Science and Technology in Trondheim. This amazing mechanism is a GPS-like tracking system which guides mammals, including us, from one location to the next[18].

Their research, conducted mostly on rodents, shows that specialized brain cells form maps of the environment that reproduce the spatial geometry of the outer world. These cells named "place" cells, located in the hippocampus, and "grid" cells, located in the adjacent entorhinal cortex (shaped like a nose), fire as the rodent moves from point to point in a certain space or enclosure. The points where the "grid" cells fire form the vertices of hexagons of successively increasing sizes that eventually fill the entire enclosure and provide a mental map to the rodent. Other specialized cells convey information to the hippocampus about the orientation of the rodent's head, its speed of displacement, and the distance to the walls of an enclosure. Note that the cells that convey speed of displacement seem to relate to movement in elapsed time.

When it comes to our perception of time, the problem is also complicated. St. Augustine, William James, Hermann von Helmholtz, Martin Heidegger, and many others agonized over it. Alan Burdick has recently written an outstanding book[19] on the subject and I am incorporating a number of his ideas here. Time is an abstract word and it can refer to several different things: a specific moment, a duration, an ordering of events, or a tense (past, present, future). As we will see, except for our consciousness of NOW, we do not have a "time organ or receptor" as we do for light and sound: we cannot perceive it directly. What we perceive is change and movement. For close to 4 billion years, the cells of all living beings on earth have contained genes that keep track of the 24 hour circadian rhythm and that program the functioning of various

18. See Scientific American, January 2016, page 28.

19. See Alan Burdick's "Why time flies," Simon & Schuster, 2017.

organs accordingly. In mammals and humans, two genes in every cell are coordinated by a so-called "central suprachiasmatic nucleus" located in the hypothalamus, just above where the optic nerves of our two eyes cross. This nucleus is triggered by sunlight impinging on us and gets reset when we travel across time zones. It acts like an internal clock but unlike a watch, it doesn't tell us the time of day and night.

Again, evolution gives us a hint as to how our brains may have developed to deal with temporal issues, including cause and effect. As we live and move around, we have to adapt from moment to moment, and our brains must guide us accordingly. This means that, contrary to what we think, we are only conscious of NOW, and then another NOW, and then the next NOW, etc. NOWs are like an endless succession of boats on a river passing through our consciousness. Many new NOWs may get stored in long-term memory immediately as they happen. They may then get retrieved from memory as a perception in a later NOW, but the past no longer exists. As for the future, it doesn't exist yet either, except that as we anticipate it and plan for it for our survival, we must imagine it NOW[20].

Burdick quotes a number of modern experiments that tell us about the properties of NOW. The perception of a flash, for example, gets to our consciousness in a fraction of a millisecond. The closest that we can resolve two separate short stimuli is 4.5 milliseconds. Events occurring under 4.5 milliseconds apart are perceived as a single event. The brain takes about 80 milliseconds to figure out what happened. A NOW awareness lasts between 5 and 15 seconds. This is, for example, why we can understand a

20. A recent book by Martin E.P. Seligman et al, "Homo Prospectus," Oxford University Press, 2016, makes the point that much of our daily mental activity is devoted to anticipating the future and developing strategies to optimize our prospects. Their point is convincing but doesn't mean that this mental preparatory activity isn't taking place NOW.

long sentence or we are sensitive to a musical rhythm, because we can mentally hold several successive words or beats in one NOW. The same is true for our awareness of cause and effect, which is necessary for survival: the tiger is here NOW, run!

When it comes to estimating duration, our brains are not very precise. The time interval during which we may be impatiently waiting at a red light is perceived as much longer than an enjoyable experience, although the durations, measured by a watch, may be the same. We may have some kind of internal pacemaker putting out pulses in our brain and accumulating (*i.e.*, counting) them, but again, it seems very subjective. The perception of duration may have something to do with how much energy is spent on a task (entropy increase), or how much information is taken in. Young children have to develop what is called *mental synchrony* to make sense of all the stimuli to which they are exposed. Until they are about 4 years old, they confuse time and speed, and have a poor sense of chronology. When we sleep, we pretty much lose whatever consciousness of time or duration we have, and we think our dreams are much longer than they really last. Without watches and knowing where the sun is, we are not very good at estimating the time of day; without calendars, we have a hard time estimating weeks, months and years; and without history books we cannot gauge how long a century is, nor verify a sequence of events as they happened.

Finally, note also that as human beings, we are uniquely able, not just to fabricate objects with our hands and machines, but also to extend the reach of our senses by means of amplifiers, lenses, microscopes, particle accelerators, mirrors, infrared goggles, telescopes, etc. so that *objects* such as nuclei, atoms, molecules, cells, objects in the dark, the planet Neptune, and distant stars are "sensed" and become part of our *representative reality*.

The Next Stage: Mental Subjective Reality

The information we receive from words, symbols, and language within a book, a conversation, a lecture, a play, a movie, the media including the so-called social media, also reaches us through our eyes or ears, but their semantic content requires memory to be decoded and identified. Words can only conjure up images, facts, actions and narratives because our brains have already stored what the words *refer to* through learning. Something similar but much more precise happens when we see or hear numbers and register them.

Two interesting questions immediately arise.

The first is: "What is the nature of all this symbolic information, and does it exist without us?" The book that lies on the table or the lecture that is recorded somewhere are independent of our existence, but their contents have no meaning without us. And what makes our species unique is that we are constantly creating a huge amount of new information. When we ask the telephone company for a new phone number, it **creates** the number and assigns it to us. The number now exists in our minds and in the telephone book. If the phone company, the phone book, and we disappear, the number ceases to have any reality.

The second question, to which we don't have a clear answer because we don't know enough yet about how our central nervous system works, is this: "How do our brains integrate all the above *sensory and symbolic representations* and combine them with knowledge already stored in our memories? Not only are we conscious of them, but we can derive opinions and concepts from them, produce histories and construct models[21].

We can discern this ability by closing our eyes, plugging up our ears and noses, etc. What should we call those experiences when we think and produce our own images with the help of our

21. The authors quoted in footnote 3 call this "model-dependent realism".

memories, or when we become aware of a state of mind (curiosity, intent, attention, expectation, determination, moral constraint) or feel emotions (happiness, sadness, fear, anger, anxiety, love, compassion, aggression or hate)? Or when we deal with abstract thoughts, numbers, mathematical formulas, purely logical structures, rules, laws, institutions and organizations, or meditations and dreams[22] that tap into our memories and our unconscious, but do not emerge from our senses? Somewhat arbitrarily, I will lump all these processes under the umbrella of *mental subjective reality*. Note that given our current lack of understanding of our brains, it is impossible to draw a clear boundary between *representative* and *mental subjective reality*. The first blends into the second in a continuous way. Think about the metaphor of the water from many rivers (our senses) running into the ocean (our brain).

As I write this essay, there are about 7.5 billion human beings on our planet. At birth the brain of a child contains an average of ~86 billion neurons[23] (roughly as many as there are stars in an average galaxy!). This number, the maximum it will ever have, slowly decays with age. Newly born babies have about 1500 interconnections (synapses) per neuron and as they grow up to adulthood and process perceptions, information, and learning, this number of synapses can increase up to 10,000 per neuron, for a total of about 100 trillion per individual. The neurons are constantly firing. If at any instant we could record the complete status of this formidable network (which is impossible), we would have the footprint of this individual's *mental subjective reality*, but not his thoughts.

22. Dreams, which I am including in this class of mental reality, will be discussed in greater detail in my later section on Psychology and Freud. Let it be noted at this point that what makes dreams special is that our brains are both their creators and spectators.

23. Chimpanzees carry 7 billion neurons in their brains, cats one billion, mice 75 million, and cockroaches one million (perhaps that is in part why they may survive a nuclear war, they have less to lose!).

Actually, what is even more significant about us humans is the enormous pool of common thoughts and knowledge we share at any given time. It's what Yuval Harari (see Ref.1 above) calls *inter-subjective reality*. It includes our various cultures, religions, beliefs and fads, languages and manners, what we learn from our families and friends and from our schools and universities, our legal and governmental institutions, our political ideas and convictions, our prejudices, our business structures, money, trade, and art. In a given society, some of this inter-subjective reality may be relatively stable but, increasingly, much of it changes continually in the entire world, particularly in this age of rapid communication through the Internet, social media, political propaganda, and commercial advertising. It is both a source of tremendous strength that allows us to use constructive ideas among large groups and organize the most complex enterprises, but also of vulnerability to the spread of dangerous movements, "fake news," hatreds, trends, influences, consumerism, and conformity.

A further problem arises for individuals with mental abnormalities or illnesses, such as paranoia or schizophrenia. For those individuals, *mental subjective reality* can become partially or totally distorted and create various forms of hallucinations or delusions. These states can be similar to what happens to people when they are dreaming, except that these individuals are awake. These states are manufactured in their heads and appear entirely *real* to them (more on this later in my section on Psychology and Freud).

Before I move on, I want to add one more comment about the visual arts, architecture, music, and what we do to embellish our environment. These activities would not exist if our perceptions did not give us the *representative realities* of shapes, colors, and sounds. But their appeal to us is highly dependent on how they tap into the part of our *mental subjective reality* that we call *aesthetics*. Although it is not universally shared, *aesthetics* is the ensemble of relations,

Fig.12 The Taj Mahal, one of the architectural wonders of the world. It is a marble mausoleum in the Indian city of Agra, commissioned in 1632 by the Mughal emperor Shah Jahan to house the tomb of Mumtaz Mahal, his favourite wife.

proportions (including *Golden ratios*), combinations, and rhythms to which our diverse cultures respond with pleasure via our conscious heritages and unconscious make-ups. When we look at a pretty flowerbed in a garden, we get pleasure. But when we look at a Monet painting of a flowerbed, the pleasure is different because the creative act of the painter taps into our much more complex *pre-existing mental subjective reality* as spectators. The *Surrealists* like Salvador Dali went one step further, trying to incorporate in their works imagined representations of the unconscious and dreams.

Fig.13 "Garden at Giverny" by Claude Monet (1840-1926).

Fig.14 "Persistence of Memory" by Salvador Dali (1904-1989).

Objective Reality?

If we accept that there are about 7.5 billion different *human mental subjective realities* on our planet at this time, does that mean that we cannot talk about such a thing as *objective reality*? Are these the unreachable *noumena* that Kant talked about? Let's think about it. From our earliest childhood, we become aware of objects outside our bodies and parts of our bodies, regardless of how we perceive them. Despite all our differences, we know that together with all living creatures we share a common world. We interact with it, we mold it and use it, produce food, tools, machines, houses, cars, roads, airplanes, and so on. We communicate with each other in it and about it. We know that the universe exists in some form, whether we are here or not to experience it and understand it. For this reason, I define *objective reality* as the **set** that includes this universe with all its components from galaxies down to elementary particles, all living species including humans, and all the objects these species create at all times, past, present, and future. Note two additional points: 1) what differentiates the elements of this gigantic and ever-changing set from each other is described by the respective properties we assign to them (admittedly through a mental process), and 2) this definition excludes *subjective* entities like thoughts, pleasures, pains, feelings, and "my consciousness" which do not exist independently of **me**; I lose these temporarily under total anesthesia, and definitively when I die.

Scientific Reality, Measurements, and Records

What makes us unique as humans is that we not only interact with *objective reality* but we also try to explore what it consists of and how it works. This is where the sciences come in. Scientific fields are commonly divided into mathematics, natural sciences (physics, astronomy, chemistry, biology, geosciences, etc.), and social sciences that study human behavior (psychology, anthropology, sociology, economics, political science, etc.). Another common characterization divides these fields into basic sciences that are mostly curiosity-driven, and applied sciences like medicine and engineering that have a utilitarian purpose. What all the basic sciences have in common is the methodology described earlier, that is used to study them: observations, theories and models invented to explain them, and experiments to verify or falsify these theories and models. The fields differ in the complexity of the systems they deal with and the resulting degree of success of their predictions.

In the next chapter I will illustrate through examples how these various sciences came about, how they are studied, and one by one, how they contribute to our quest for reality. Because of the almost infinite scope of today's scientific enterprise, my examples will obviously not be exhaustive. I must also warn the reader that my choices may appear somewhat arbitrary.

PART 3:

MATHEMATICS AND THE NATURAL AND SOCIAL SCIENCES

Heads up: when we refer to an entity, an object or a process, we generally define it through some characteristic properties. Scientists consider this entity "real" if they can make records or measurements of these properties by means of instruments[24].

A measurement may yield a number or a function of numbers for the magnitude of a property, the location of an entity, the direction of an effect, the severity of a disease or the degree of a physiological impairment, and so on. The beauty of numbers is that educated people can all understand their values.

The same rule is true in a court of law where some kind of record must exist and be found to establish that a fact or event is "real".

Mathematics and Computer Languages

The creation of mathematics is one of the miracles of the human mind. Mathematics is a universal language that has many branches such as arithmetic, algebra, geometry, trigonometry, calculus, statistics, number theory, group theory, etc. The Greeks made numerous early contributions to the study of mathematics, and, not by coincidence, its foundations are reminiscent of the syllogisms described earlier. Modern mathematics is based on only nine axioms from which all rules can be derived. Axioms are absolute statements that cannot be proved but can lead to provable theorems. Whatever mathematical knowledge we acquire, it becomes a part of our *mental reality*. What is so amazing is how rules, once established, can be extended. In algebra for example, when we define certain categories of numbers (integers, negative numbers, rational numbers, irrational numbers, real numbers), we discover that they obey certain rules such as addition, subtraction, multiplication, division, etc. Mathematicians can then extend these rules to invent new categories of

24. See article in "Reference Frame" by N. David Mermin in Physics Today, May 2009, and answers by Michael Nauenberg and colleagues in letters in Physics Today, September 2009.

numbers such as imaginary (!) and complex numbers which then spill over into geometry and trigonometry[25].

In its application, does mathematics liberate physics and other sciences from the limitations of written language? To answer this question, we have to examine what mathematics does. As mentioned earlier, one property of mathematics that makes it much more powerful and more flexible than common language is that the symbols on which it operates do not have a semantic content per se. Essentially all physical laws can be expressed by mathematical formulas, equations, and functions, whether they involve gravity, mechanics, electricity, magnetism, quantum mechanics, nuclear phenomena, energy or heat flow, astronomy, etc. Functions of various variables can be manipulated and equated to other functions of other variables; the functions can be differentiated or integrated; the variables can be extrapolated to zero or to infinity; numbers can be "real" or "complex"; geometries can be Euclidean (representing flat space) or "curved" (representing space-time); and three-dimensional space can be generalized to ten-dimensional space, just to give a few examples.

When physicists want to study the behavior of a system in space-time, they can start with assumptions about this system including "initial conditions," assign names or symbols to its properties, and incorporate these into equations such as Newton's second law which states, "force equals mass times acceleration."

For example, if we drop a lead ball from the tower of Pisa, we can manipulate Newton's equations following the strict rules of calculus, and make exact predictions of when the ball will hit the ground if we know the height of the tower. If instead of a lead ball, we drop a sheet of paper, things get much more complicated: we will have to know the exact mass and shape of the sheet, the direction

25. Richard Feynman masterfully illustrated this fact in his chapter 22 on Algebra in his "Lectures on Physics," Addison-Wesley Publishing Company, Inc., 1963 with Euler's formula $e^{i\Theta} = \cos\Theta + i\sin\Theta$.

and speed of the wind at all times and levels, etc. and it will take a very complicated computer program to predict when and where the sheet will land. However, we may still be successful in making predictions that match experimental results[26], without using any words. This is part of the process that kept Einstein in awe. It is also this very process of manipulating formulas and functions that led him to his famous equation $E_0=mc^2$, which states that the energy of an object at rest is equal to its mass multiplied by the square of the speed of light[27].

There is another powerful aspect of mathematics to which Emmy Noether (1882-1935), a German mathematician much admired by Einstein, made a pioneering contribution. It has helped physicists discover laws of "conservation" of certain quantities like "energy of a system," "angular momentum," and "electric charge." "Conservation" means that these quantities remain constant when certain operations like translations, rotations, etc. are applied. These operations are broadly called "symmetries" of which the most common is the reflection symmetry produced by a mirror. Spoken or written language could never make these discoveries.

With the help of computer programs, it is possible to model and predict the behavior of increasingly more complex systems with which simple calculus and equations cannot cope, like for example the above falling of the sheet of paper, or how in 1969 the Apollo mission got to the moon and back. Note that the full computing

26. Note that using a law like Newton's second law in combination with calculus describes a deterministic process. For much more complex systems, like predicting the weather, initial conditions are never completely known and small changes in these input conditions can lead to vastly different outcomes. In such cases, we are still dealing with a deterministic process but the outcomes are called "chaotic". The theory of chaos that deals with these systems has wide applications in many fields other than physics.

27. Note that Einstein's famous equation is most often written simply as $E = mc^2$ without the subscript 0, but this latter form is not the one he derived originally, and it can lead to confusion because E refers to total energy, i.e., rest energy plus kinetic energy (due to motion).

power the Apollo mission had on board at the time is now contained in a single smartphone!

Fig.15 German mathematician Emmy Noether (1882-1935) who helped discover "conservation laws" in physics.

Many of the operations described here are so abstract that they can stretch our world view beyond our imagination without violating any of the axioms from which they are derived. On the other hand, it seems to me that mathematics or computer programs that can be used to derive and solve equations relating certain quantities to other quantities cannot "invent" concepts like energy, mass, or momentum. These are created separately by our minds, our imagination and our observations, before their values are "plugged" into these equations.

The Natural Sciences

Physics and Astrophysics

The physical world at our macroscopic scale is described successfully by the branches of classical physics: classical mechanics, thermodynamics and statistical mechanics, electromagnetism, optics, hydrodynamics, plasma and solid state physics. At the astrophysical and microscopic scales the world is currently described by two very successful yet incomplete so-called "standard" models. The first is the Standard Model of Cosmology and the second is the Standard Model of Particle Physics. For a summary of where these models stand today, see Appendix 1. The models are joined at the hip by two theories that underpin them. One is Einstein's Theory of Relativity which is deterministic. For readers with some mathematical training that may be interested in Einstein's Time Dilation and Equivalence Principle, see Appendix 2. The other theory is Quantum Mechanics and Field Theory which is ruled by "wave functions" or "probability amplitudes" which yield probabilities of finding the fields and particles in certain conditions at certain times[28]. For details of how the ubiquitous photon or quantum of light came about, see Appendix 3.

To illustrate the contrasts between classical and quantum physics, let us start with the simple example of a spherical stone and examine a set of its properties: its shape, its color (the light wavelengths it reflects), its hardness, its constituents, its mass and its weight, its position and its momentum. We can record the exact shape of the stone with a picture and measure its other features. Of course, we must be careful to define the properties we measure. The

28 For a discussion of the advent of quantum mechanics in the 20th century, see for example A. Douglas Stone's "Einstein and the Quantum," Princeton University Press, 2013.

mass of the stone is defined as an intrinsic property and depends on the number of atoms it contains. It doesn't change, whether we observe it or not. Its weight, on the other hand, depends on where it is measured: it is much lighter on the moon than on earth because of the difference in the respective gravitational fields, and it is zero on the space station. If we catapult the stone from point A to a distant point B, we can predict its trajectory to great accuracy by using classical mechanics. We can also track it in transit by bouncing visible light or radar signals (photons) off it, without disturbing it much, so the position of the stone can be known at all times.

If we drop our stone into a quiet pond (a "field" of water), it will create ripples on the surface of the water that will travel outward as circular waves. What causes the waves is that the stone creates a temporary void in the water while it sinks, and the water swishes back and forth into the void like a pendulum, making the level of the surface oscillate. The bigger the stone, the bigger the outgoing ripples, but not the distance between them (wavelength) which is a property of the water. These ripples are real, they can be recorded; their amplitudes, speed of propagation and wavelength can be measured. The picture is fairly simple and clear.

Fig.16 Ripples on a water pond produced by a falling stone at the center.

However, moving on to the physics of fields and elementary particles, we are presented with a subtler ontological and epistemological challenge. On a macroscopic scale, we can agree on the existence of the stone and our knowledge of its properties discussed above. But what about an elementary particle like the ubiquitous electron? As far as we know today, its existence and behavior are ruled by quantum mechanics and its wave functions. Furthermore, quantum field theory (QFT) states that logically the electric field comes first and that the electron is a ripple excited in it. Once it manifests itself, the ubiquitous electron, like the above stone, can be characterized by a set of inferred properties: its mass, its spin, its electric charge, and its position and momentum; it can be found bound in orbital locations around the nucleus of an atom; it can also be pried loose and move by itself or with other electrons in a beam through a vacuum or a metal where its flow results in an electric current and an associated magnetic field around it. However, upon closer scrutiny, physicists believe that the electron really exists in two massless quantum states with opposite spins, the actual direction of which only becomes determined when measured, that the mass of the electron is probably acquired through interaction with the Higgs field, that its observed electrical charge is not its real "bare" charge which is masked by surrounding pairs of electrons and positrons fading in and out of existence, and that under certain conditions the electron can change energy by either absorbing or emitting a photon of light. The experimental and theoretical evidence formulated by Heisenberg's Uncertainty Principle tells us that we cannot simultaneously measure the electron's position and momentum to full accuracy. According to the Dirac equation, the electron has an anti-particle, the positron, with the same electric charge but of opposite (positive) sign.

Fig.17 Albert Einstein (1879-1955).

No sooner was quantum mechanics born around 1925, than its implications began to generate endless philosophical discussions between Niels Bohr, Werner Heisenberg, Max Born, John von Neumann, Wolfgang Pauli and other contemporaries on one side (the camp with the so-called "Copenhagen interpretation"), and Albert Einstein and to some extent Erwin Schrodinger on the other.

Fig.18 Werner Heisenberg and Niels Bohr, probably in a friendly discussion about quantum mechanics in Copenhagen, sometime before WW2 when political circumstances complicated their relationship.

Where did the major disagreements come from? There were probably three major points of contention. The first was the probabilistic nature of the predictions made by the wave functions. Einstein, despite his important contributions to the development of quantum mechanics, felt intuitively that nature was ruled by deterministic laws and that quantum mechanics was missing something: *i.e.*, it was not complete, there had to be some "hidden variables" or extra properties not appearing in the equations. The second was that the Copenhagen camp held that the properties of a particle are undefined (*i.e.*, don't exist or aren't real) until they are measured, at which instant the "fuzzy" wave function that represents it collapses at a single point and the particle that appears becomes real. Bohr referred to these properties, wave-like and particle-like, as being *complementary* but not simultaneous. Einstein could not accept this idea that seemed to imply that *objective reality* depends on the act of measuring it, somewhat reminiscent of 18th century Bishop George Berkeley's view of reality. The third point, related to the second one, had to do with Einstein's belief in the principle of "locality" or the impossibility of "spooky action at a distance." What he meant was that if some physical phenomenon takes place at a given point, it has to be triggered by the local presence of a particle or a field at that point. Furthermore, this particle or field cannot get there instantly from somewhere else, faster than the speed of light, because this would violate his Theory of Special Relativity. None of these points seemed to deter the Copenhagen camp, however!

Fig.19 Irish English philosopher Bishop George Berkeley (1685-1753) whose central idea was that "to be is to be perceived." The city of Berkeley where the University of California is located is named after him.

Almost a century later, these controversies are still not settled[29] despite the fact that quantum mechanics and field theory have been highly successful in predicting (but not fully explaining) many observations and enabling endless technological innovations. Several physicists have tried to find ways around the above difficulties. David Bohm (1917-1992) came up in 1952 with a "pilot-wave" theory that, similarly to Louis de Broglie's, associated a guiding wave to the "real" particle that determined its trajectory unequivocally. Bohm's idea eventually gained adherents but was dismissed by most of the Copenhagen camp as contrived or unnecessary. Hugh Everett III (1930-1982) in his Ph.D. thesis at Princeton in 1956 got around the idea that the wave function needed to collapse by assuming that there is only one gigantic wave function that describes the entire universe and the quantum states of all events: it just "branches out" into alternate worlds, every time something happens. The idea came to be known as the "many-world" theory. Although admired by a few contemporary physicists such as John Wheeler (Everett's thesis advisor), it was dismissed by most because of its complexity: indeed no wave function collapse, but an infinite number of worlds! The next, and perhaps most influential, actor in this saga was John S. Bell (1928-1990). In 1964, Bell came up with a mathematical inequality that could be experimentally verifiable or falsifiable (à la Karl Popper). If the inequality was correct, then nature could be local and quantum mechanics is incomplete; if the inequality was not correct, then quantum mechanics is OK and "spooky action at a distance" is possible. Today, fifty years later, the answers gotten from a number of experiments are apparently still not definitive. But in the meantime, a number of experimental physicists have observed so-called "entangled" pairs of particles like electrons and positrons or photons that, when created at one point,

29. For an up-to-date discussion of this subject, see Adam Becker, "What is real? The unfinished quest for the meaning of quantum physics," Basic Books, 2018.

would share the same wave function and are able to communicate **instantaneously** after they fly far apart[30]. Non-locality, *i.e.*, action at a distance, would thus be possible.

Fig.20 Richard Feynman (1918-1988) teaching quantum mechanics at Caltech.

Meanwhile, some contemporary theorists want to "reify" the quantum wave functions (which I would claim are parts of our *mental reality*) as if they were physical objects. They consider quantum mechanics as the "fundamental theory" and view classical mechanics as an approximation valid to explain macroscopic-scale phenomena. As Richard Feynman (1918-1988), the famous Caltech physicist who made major contributions to quantum mechanics, said: "I think I can safely say nobody understands Quantum Mechanics." As unorthodox as it sounds today, Einstein may have been right in his objections after all!

Recently, gravitational waves theoretically predicted by Einstein in 1916 have finally gained the status of *scientific reality* through five consecutive observations. The first were observed on September

30. See for example article on "Reality Check" by Michael Brooks, New Scientist, Volume 219 No. 2928, August 3-9, 2013, page 33.

14th, 2015, by the LIGO observatories (two 4 km-long advanced laser interferometers, one in Hanford, WA, the other in Livingston, LA) and released officially on February 11th, 2016. These first observations are ripples in space-time that were generated 1.2 billion years ago when two black holes of respectively 36 and 29 solar masses, rotating around each other, finally collided and merged into a single black hole, releasing energy equivalent to roughly three solar masses. The black holes represented as white balls, the waves they generated, and the recordings of the miniscule mechanical strains they produced on the two detectors 1.2 billion light-years away are shown below.

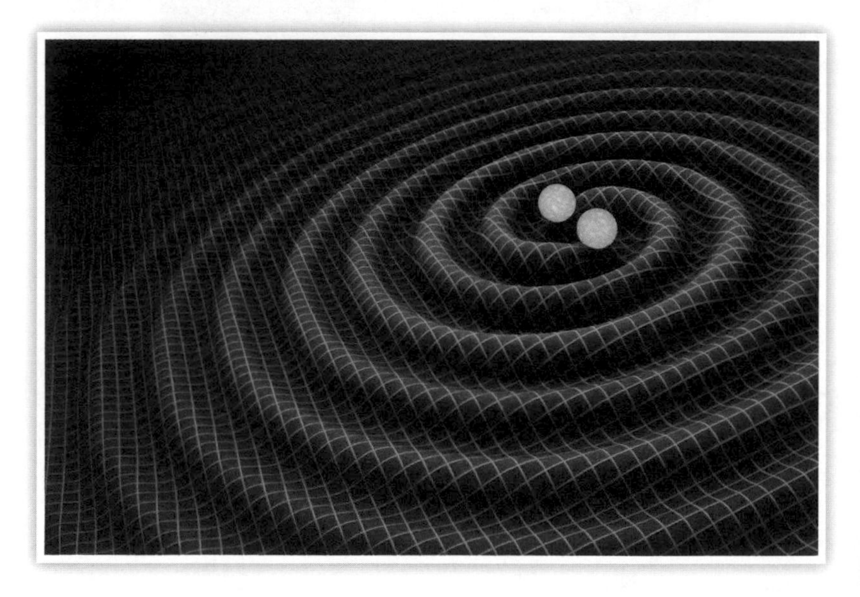

Fig.21 Ripples in space-time produced by two merging black holes 1.2 billion years ago.

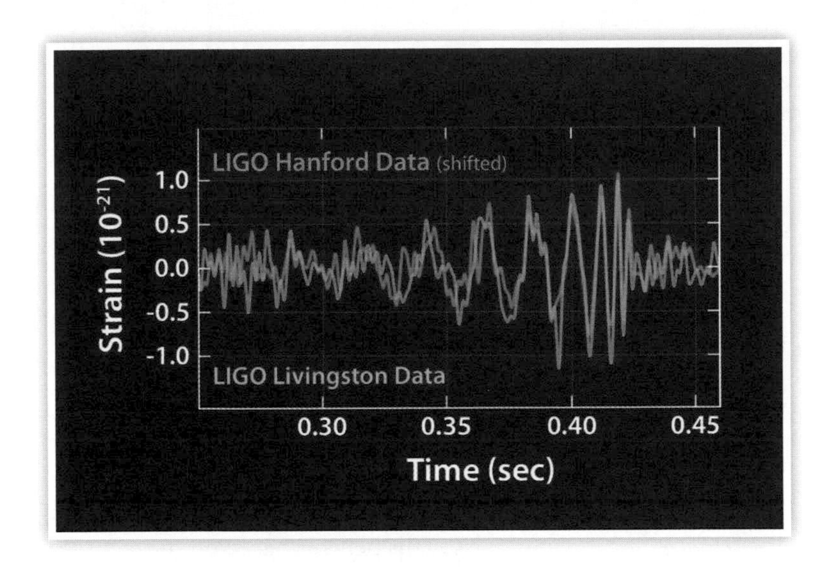

Fig.22 Gravitational waves resulting from the above merger detected by the two LIGO interferometers in 2016.

Subsequently, three similar events resulting from other pairs of black holes merging elsewhere in the universe were observed during the next year. More recently, on August 17th, 2017, the two LIGO instruments together with the similar European Virgo instrument observed a fifth gravitational wave event from a merger of two neutron stars of a combined mass of ~2.75 solar masses. This collision, unlike that of black holes, simultaneously produced electromagnetic waves observed as gamma rays by the Fermi space telescope and in other bands of the electromagnetic spectrum by as many as 70 other ground-based telescopes and observatories all over the world. These five observations are just the beginning of an extremely promising new branch of astrophysics. Not surprisingly, the 2017 Nobel Prize in Physics was awarded to Rainer Weiss, Kip Thorne, and Barry Barish for their seminal role in these discoveries.

Chemistry

Let us now move to the field of chemistry. It offers another perfect example of how a discipline's evolution has informed our "quest for reality."

In antiquity, Indians, Chinese, Egyptians, Babylonians, and Greeks all learned to handle materials to build shelters, make tools, pottery, weapons, etc., but didn't do many experiments to identify the materials' fundamental ingredients. Democritus (and probably his teacher Leucippus) believed that everything in nature is composed of "atoms" that are physically indivisible, eternal, and separated by empty space. The word "atom" stems from a Greek word meaning "uncuttable." Democritus of course had no experimental proof of this but was probably guided by a belief in *reductionism*, which is still fashionable today. This belief is based on the intuition that objects can be divided into ever smaller sub-parts with similar properties, down to a limiting point. Aristotle (384-322 BCE), pupil of Plato and tutor of Alexander the Great, didn't believe in this idea and held that the world is made of four elements: air, water, earth, and fire. Because of his enormous intellectual influence, Aristotle put the idea of atoms to bed for two thousand years, until new observations supporting the concept were finally made in the 19th century.

Following the work of Lavoisier and Dalton, Russian chemist Dmitri Mendeleev (1834-1907) must be recognized for his seminal effort to bring order into the properties of all roughly 60 then-known atomic elements and to summarize them in a presentation to the Russian Chemical Society on March 6, 1869, entitled "The Dependence between the Properties of the Atomic Weights of the Elements." The table was subsequently republished in the German journal *Zeitschrift Fur Chemie* and became the foundation of today's Periodic Table of Elements. It lists the elements according to their atomic numbers (numbers of protons or electrons) rather than their atomic weights, not shown here (which include the masses of the neutrons in addition to the protons that their nuclei contain). For example, the element carbon has the atomic number "6" but exists in nature with 6, 7, or 8 neutrons that give them atomic weights of 12, 13, or 14 respectively. Carbon-14 is radioactive and its decay is used to date the age of organic substances. What is so amazing about these tables is that their arrangement in columns, which corresponded to the chemical valences of the elements, predicted when an element was missing and led to its subsequent discovery. In the same way Darwin didn't know about genetics, Mendeleev didn't know about protons, neutrons, or electrons whose orbital configurations in atoms and molecules today can predict much of their chemical behavior.

1 H																		2 He
3 Li	4 Be												5 B	6 C	7 N	8 O	9 F	10 Ne
11 Na	12 Mg												13 Al	14 Si	15 P	16 S	17 Cl	18 Ar
19 K	20 Ca		21 Sc	22 Ti	23 V	24 Cr	25 Mn	26 Fe	27 Co	28 Ni	29 Cu	30 Zn	31 Ga	32 Ge	33 As	34 Se	35 Br	36 Kr
37 Rb	38 Sr		39 Y	40 Zr	41 Nb	42 Mo	43 Tc	44 Ru	45 Rh	46 Pd	47 Ag	48 Cd	49 In	50 Sn	51 Sb	52 Te	53 I	54 Xe
55 Cs	56 Ba	*	71 Lu	72 Hf	73 Ta	74 W	75 Re	76 Os	77 Ir	78 Pt	79 Au	80 Hg	81 Tl	82 Pb	83 Bi	84 Po	85 At	86 Rn
87 Fr	88 Ra	**	103 Lr	104 Rf	105 Db	106 Sg	107 Bh	108 Hs	109 Mt	110 Ds	111 Rg	112 Cn	113 Nh	114 Fl	115 Mc	116 Lv	117 Ts	118 Og
		*	57 La	58 Ce	59 Pr	60 Nd	61 Pm	62 Sm	63 Eu	64 Gd	65 Tb	66 Dy	67 Ho	68 Er	69 Tm	70 Yb		
		**	89 Ac	90 Th	91 Pa	92 U	93 Np	94 Pu	95 Am	96 Cm	97 Bk	98 Cf	99 Es	100 Fm	101 Md	102 No		

Fig.23 Periodic Table of the Elements.

Elements like Radium (88) and Polonium (84) were discovered by Marie and Pierre Curie, who found them to be *radioactive, i.e.,* to decay spontaneously into other atoms by emitting radiation. Democritus turned out to be wrong in this respect, although he of course didn't know about quarks and leptons, the next layer of physical reality which may be "uncuttable." For their discovery, the Curies shared the 1903 Nobel Prize with Henri Becquerel, who had discovered radioactive uranium (92) salts. Many other radioactive elements were discovered in subsequent years.

Note that it took more than another century to understand where all these elements were created. The explanation now turns out to be that the elements came into being in probably three astonishing separate steps. The first two elements, hydrogen and helium, plus small quantities of lithium, beryllium, and boron were synthesized in our universe three minutes after the Big Bang. The next elements, from carbon up to iron, were synthesized much later in low-mass stars and during the collapse of massive supernovae, without which we wouldn't exist. As of 2017, the higher mass elements beyond iron were probably created during the collisions of neutron stars!

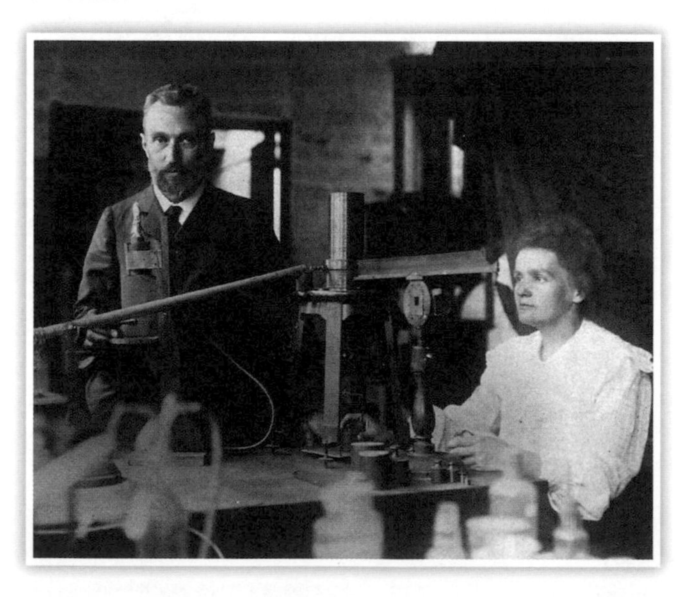

Fig.24 Marie Sklodowska Curie (1867-1934) and Pierre Curie (1859-1906) who discovered two radioactive substances, radium and polonium.

Originally the word "chemistry" came from "alchemy," the Arabic *al-kimiya* and the late Greek *chemeia* which stood for black magic. Alchemy, or black magic, was unscientific, but its semi-secret practice went on for centuries. Its never realized goal, among others, was to transform abundant elements like lead into the more rare and precious element of gold.

By the middle of the 20th century, "chemistry" came to mean the science of substances: their structure, their properties, and the reactions that change them into other substances. However, for philosophical reasons not based on any scientific ground, its study was divided into inorganic and organic chemistry. The idea was that organic chemistry was reserved to the study of substances containing carbon, such as hydrocarbons and compounds derived from living organisms. The study of the latter was believed to belong to a separate science because it was ruled by an additional principle, the *elan vital* (vital force) not present in inorganic compounds. Although this belief is no longer held today, the division still exists. Organic chemistry is the bridge to biochemistry and biology.

Biology and Genetics

While *elan vital* was never found in biological substances, their ability to self-organize, consume energy, store information and reproduce was the source of a great puzzle: where does it come from and how does it work? After decades of research by many investigators, a huge leap forward was made in 1953 through the discovery of the DNA double-helix, a structure joined by transverse rungs of base pairs whose sequence expresses a genetic code (see below). The discovery was the achievement of five scientists: James Watson (1928-), Francis Crick (1916-2004), Maurice Wilkins (1916-2004), Rosalind Franklin (1920-1958), and Raymond Gosling

(1926-2015). Watson and Crick, working in Sir Lawrence Bragg's Cambridge Cavendish Laboratory, were not experimentalists and alone could not have identified the DNA structure correctly. They obtained the various DNA x-ray diffraction pictures, *i.e.*, the "reality check," from Wilkins, Franklin, and Gosling at London's Kings College but didn't give them the proper credit. The controversial circumstances underlying these events were not immediately known and the Nobel Committee awarded Watson, Crick, and Wilkins the Nobel Prize for Physiology or Medicine in 1962 for the discovery. Franklin and Gosling's findings were eventually vindicated and their contributions were fully recognized. Sadly, Franklin had died in 1958 before she could learn of her recognition.

Fig.25 Swedish postage stamp commemorating the 1962 Nobel Prize for Physiology or Medicine given to Crick, Watson and Wilkins for the discovery of the double-helix structure of DNA.

Fig.26 Rosalind E. Franklin (1920-1958), English chemist who made seminal contributions to the understanding of the structure of DNA at King's College in London.

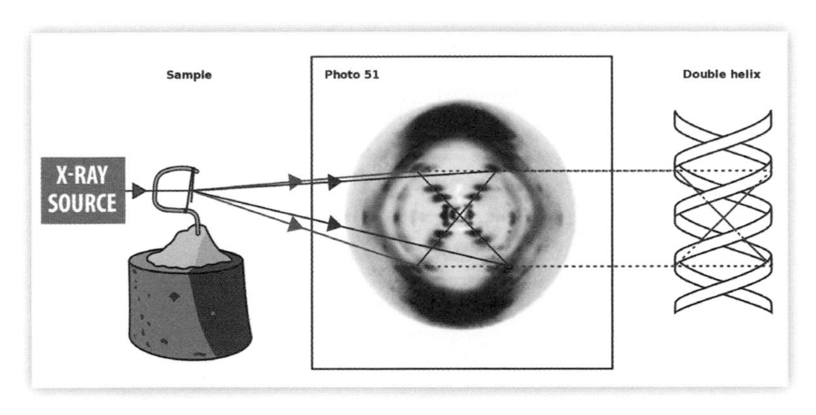

Fig.27 Photo 51 was an X-ray image of a DNA sample taken by Raymond Gosling, doctoral student of Rosalind Franklin. It was from this image that she figured out the double-helix structure.

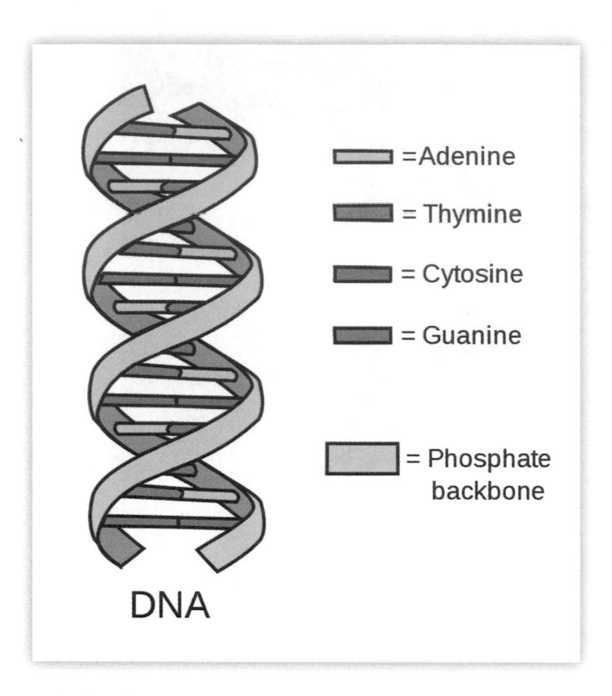

Fig.28 DNA double-helix structure showing helical sugar phosphate backbones and transverse base-pair rungs (AT and GC) whose different sequences code for different proteins (amino-acid chains). The rungs are 3.47 Angstroms apart.

Huge progress has been made in this field since 1953. Here, in broad lines, is a very short summary of the fundamental advances:

A grown human has about 37 trillion cells of various types and functions. These cells contain a nucleus separated by partially porous membranes from a surrounding cellular medium called cytoplasm. Except for the mitochondria (organelles producing energy in the cell), the nucleus contains all the genetic material in the form of long DNA chains which make up the human genome. The genome contains the chromosomes which in turn contain the individual gene coding for specific traits. The total length of the human genome includes over 3 billion base pairs (made up of the paired nucleotides, Adenine-Thymine and Guanine-Cytosine, as shown in the figure). It was first fully sequenced in 2001. There are 22 pairs of chromosomes which are the same for males and females. The 23rd pair of sex chromosomes differs between males and females. Females have two copies of the X chromosome, while males have one X and one Y chromosome. This makes for a total of 46 chromosomes per individual.

The machinery of cell multiplication through division is now fairly well understood. The DNA in a cell, before the cell splits (through *mitosis* for so-called somatic, *i.e.*, non-sexual cells, or *meiosis* for sexual cells), is replicated by RNA, a molecule similar in many ways to DNA but single-stranded. DNA and RNA are both nucleic acids which, along with fats, proteins, and carbohydrates make up the macromolecules essential for life. The RNA molecule synthesizes the new DNA and proteins. Every one of the myriad of different proteins making up our bodies is a chain of various amino acids (twenty in total). It takes three paired nucleotides to code for each amino acid that the RNA then assembles into proteins in a structure called ribosome. There are an estimated 21,000 human protein-coding genes in our genome, but the fraction of each gene sequence responsible for the coding is only about 1.5%. The rest is associated with non-coding RNA molecules and many other sections whose functions have not been fully determined. Maybe the *elan vital* is coded in there somewhere!

According to new discoveries made in a field called Epigenetics[31], if an organism is subjected to certain environmental conditions like stress, famine or strong trauma such as maternal separation, its DNA remains unchanged but the manner in which it is transcribed may be altered. This alteration can result in long-lasting changes in how specific genes are expressed or shut-down, and can be transmitted to an off-spring in the womb. An example of this

31. See Israel Rosenfield and Edward Ziff in "Epigenetics: The Evolution Revolution", The New York Review of Books, June 7th, 2018, page 36.

effect is that the receptors in an organism that inhibit the production of a certain hormone preparing it for stress are deactivated. The uninterrupted production of this hormone (even in the absence of stress) can result in heart disease, diabetes, inflammation, schizophrenia and depression. Whether such an effect can be transmitted to future generations is not yet established.

I will come back in Part 5 to major developments in the field of genetics that are spawning revolutionary and controversial technologies such as GMOs and CRISPR.

Medicine and Healthcare

Turning to medicine, it is often asked whether it is a full-fledged science. Indeed, like biology, medicine uses scientific methods to study human anatomy and physiology. However, when we go to the doctor, our main purpose is not to study our body but to find out how to prevent or diagnose and treat a disease, accident, or abnormality. Because of the extreme complexity of the human body and the fact that the doctor has to use intuition to choose among multiple possibilities, this process is more like solving a detective story. Theories can be developed but experiments are much harder to perform. Prediction often doesn't quite work. The human body is constantly changing under various influences. Anatomy is well known in general but the anatomy of every human being differs. There are still many variations among humans that are not measurable. Physiology, where everything interacts, is even more complex. Certain conditions may result in no symptoms in one individual but in constant pain in another. Diagnosing the cause of an illness can be fairly simple in certain cases and almost impossible in others. Even assuming that a health problem has been correctly diagnosed, treatments may not always be available and if they are, many of them are still based on trial-and-error methods: "Take this pill with lots of water twice a day, come back in a week, and we'll see how you feel." And pills often have undesirable side effects.

Having said this, medicine is making amazing progress in many areas. Rapid advances are being achieved in testing techniques, imaging instrumentation, drug design, vaccines, surgical techniques, radiotherapy, chemotherapy, immunotherapy, nutrition research, predictions and remedies based on genetics, regenerative medicine through stem cell treatments, mental health, various physical therapies, and in other general areas such as record keeping. Some futurologists predict that pattern recognition machines combined with large data banks may someday come up with better and faster diagnoses than doctors do today. This is possible, but I don't think medical doctors will be replaced by machines anytime soon!

Meanwhile, on a world scale, viral and bacterial infections with their resulting pandemics still present daunting challenges: just think about HIV and the recent Ebola outbreaks in Western Africa and the Congo! They are unpredictable and require constant vigilance. The necessary vaccines are not always available and can be very costly. Even in an advanced country like the U.S., the fact that about one third of the adult population is obese and 10% suffer from diabetes, that millions of people are addicted to opioids, that autoimmune disorders are not curable, and that allergies and chronic disorders are rampant means that medicine still has huge challenges ahead.

And then there is the enormous problem of healthcare cost and insurance. As this book is being written, a number of OECD countries (like France, the U.K., Canada, Spain, Germany, Italy, Australia, and Japan) have found ways of handling this problem fairly efficiently and equitably across social classes. They spend the equivalent of about $6000 (give or take 10%) per capita per year for healthcare. This is to be compared with the roughly $10,000 per capita per year in the U.S. where politicians are having a very hard time creating an efficient and equitable healthcare funding and insurance system. At the other extreme, there are probably over three billion people in the developing world for which countries are spending much less than $1000 per capita per year on the healthcare of their populations. As they find out that they are being deprived of care that exists for others, they will become increasingly unhappy. Further population growth and climate change will put enormous stresses on the entire medical system in the future.

Geosciences, Energy, and the Environment

Let's start with the earth. Although Ancient Greeks often proposed theories without being able to test them, this was certainly not the case for Eratosthenes (c.a. 276-194 BCE). Eratosthenes was a remarkable Greek mathematician, astronomer, geographer and poet born in Cyrene, Libya and deceased in Alexandria, Egypt. Following the ideas left behind by Anaxagoras of Clazomenae (c.a. 500-429 BCE) and Aristarchus of Samos (c.a. 310-230 BCE), and his own observations, he knew that the earth was a sphere and he was the first to calculate its circumference. He had heard of a famous well in Syene, near modern-day Aswan, where every June 21st (day of the summer solstice), the sun directly overhead at noon would shine its rays straight down to the bottom, along the vertical in the direction of the center of the earth and would be reflected straight up by the water. At the same instant, in Alexandria, several hundred kilometers to the north, the sun would hit the ground at an angle off the vertical which he was able to measure as 7.2 degrees or 7.2/360 or 1/50th of the earth circumference. From a simple geometrical sketch, he knew this angle to be the same as the angle of the vertical from Alexandria to the center of the earth with respect to the vertical in Syene. From Pharaonic records (or measurements he may himself have ordered) he also knew that the distance between Syene and Alexandria (*i.e.*, the length of the arc on earth) was 5000 Olympic stadia (or stades), a unit of length which in Egypt corresponded to today's 157 meters. Thus the circumference of the spherical earth was:

$$50 \times 5000 \times .157 = 39{,}250 \text{ km}$$

(accurate to within 2% of today's measurement). Amazing!

Following this feat, Eratosthenes also figured out the sizes and distances from earth of the moon and the sun[32]. As a true scientist, he produced numerical records of reality! Eratosthenes is also considered the father of modern geography. He left a map of the world as it was known at the time and became the chief librarian of the Library of Alexandria. Unfortunately, many of his ideas were mistrusted and forgotten for a long time.

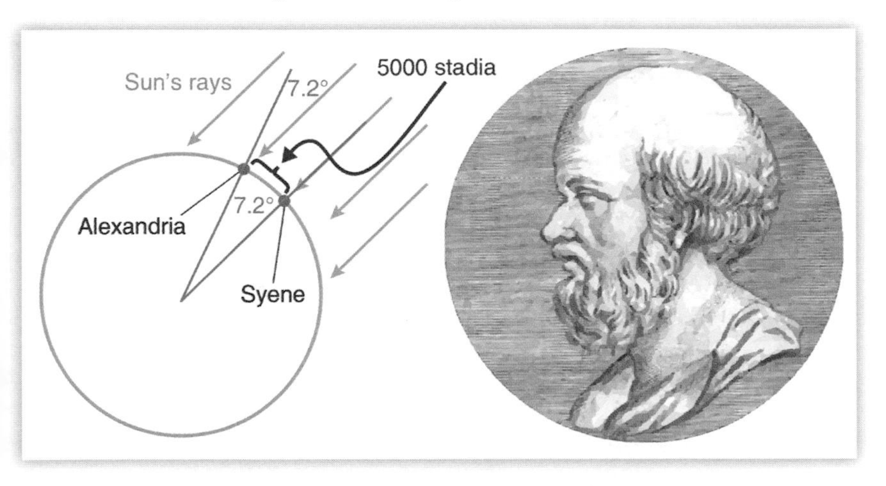

Fig.29 Eratosthenes (~276-194 BCE), Greek mathematician, astronomer, geographer and poet, who was born in Libya and lived in Egypt where he calculated the circumference of the earth through an incredibly ingenious physical measurement.

32. See Simon Singh, "Big Bang," Harper Collins Publishers Inc., 2004, pages 11-18.

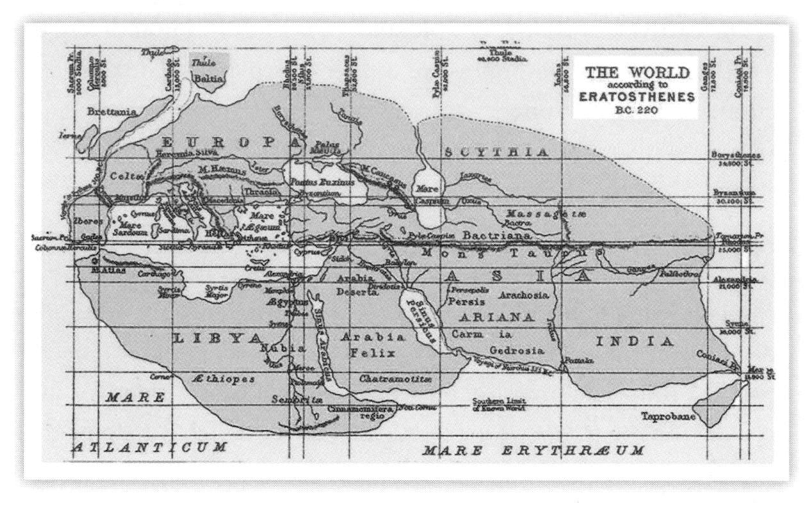

Fig.30 Eratosthenes' map of the earth as it was known at the time, showing parallels and meridians for the first time.

Now, fast forward by more than two millennia. Like Eratosthenes, you find armies of geoscientists who not only know mathematics but have new tools at their disposal from physics, chemistry, and biology. They study what the earth consists of, how its various systems (land, oceans, forests, deserts, tundra, polar ice, magnetic field, atmosphere, ionosphere, troposphere, etc.) work, how they evolve, and how we can benefit from their resources or harm them.

Our planet is about 4.6 billion years old and it is complex and diverse. Predictions about some of its behavior are still hard to make. We know about the earth's crust (the lithosphere), its mantle, and to some extent about its liquid and solid, hot and radioactive core (hard to measure). We know about the 7-8 major tectonic plates in the lithosphere and upper mantle which slowly drift around under the effects of various forces and thereby reconfigure the shapes of our continents. We know that volcanoes and earthquakes appear predominantly at the boundaries of these plates but we cannot yet accurately predict the eruptions of volcanoes and the occurrences of earthquakes. We know a lot about climate but we can't predict the weather two weeks ahead.

From our selfish point of view, our planet is at the right distance from the sun and has enough heat in its core to maintain a reasonable temperature without very large seasonal fluctuations, and with the availability of water. These conditions make life as we know it possible. We tend to think that animals are superior to plants and trees, but from an energetic point of view, the latter are much more self-sufficient in that they can't run around and must manufacture their food *in situ*, using the miracle of photosynthesis.

Photosynthesis uses energy from sunlight that falls on leaves or needles, which act like "lungs." Chlorophyll in the leaves or needles absorbs light energy and transfers it to chemical reactions to absorb the CO_2 from the air, break up the water molecules (primarily absorbed by and rising from the roots) into hydrogen and oxygen, make sugar in the form of glucose ($C_6H_{12}O_6$) and release oxygen to the air. Much of the sugar, as in our bodies, temporarily stores the energy coming from the sun, circulates in the tree, performs its functions, and in the process of delivering its energy, makes CO_2 again, which is *expired, i.e.*, re-emitted to the atmosphere, like when we exhale. In most trees, the CO_2 intake dominates during the day when there is sunlight, and the CO_2 release dominates (but is less) during the night. In winter, for deciduous trees (*i.e.*, that shed their leaves), the photosynthesis process stops or is limited to green twigs. Mostly in the summer, trees also release a large amount of water through *transpiration* and *evaporation*.

About one third of our planet's land, or 39 million square kilometers, is still covered with natural forests that do not require irrigation or man-made fertilizers or energy other than what they receive from the sun. They are essentially self-sustaining, except when there are forest fires which require firefighting. Wild animals that eat wild plants or predate on other animals can also be said to be self-sustaining from an energy point of view. On the other hand, plants planted by humans for agriculture or decoration require extra water, fertilizers, and increasing energy. Animals domesticated for milk and meat production are also not self-sustaining. Cattle exhale large amount of methane. Livestock contribute over 5% of all greenhouse gases. Humans are by far the greatest consumers of water, raw materials, and energy on earth. Our annual world energy consumption is about 160,000 TWh (one Terawatt-hour, or TWh, is equal to one billion kilowatt-hours) and is still growing globally by about 2% per year. More than 80% of the energy comes from fossil fuels, 10% from nuclear energy, and the rest from hydropower, biofuels, and renewable sources like solar and wind. Note that fossil fuels originally also came from solar energy in the form of decayed wood and bacteria stored in the ground over millions of years. Unfortunately for us, when we burn them, they release greenhouse gases into the atmosphere which trap heat and gradually warm up the earth.

Our atmosphere contains on average, by volume, 78% nitrogen, 21% oxygen, water vapor (highly variable, 0.01-3%), 0.9% argon, 0.04% carbon dioxide, 0.0002% methane (CH_4) (not accurately measured, and may be higher), and smaller quantities of nitrous oxide (NO_2), various halocarbons (containing fluorine), ozone (O_3), neon, and helium. Water vapor, carbon dioxide, methane, nitrous oxide, ozone, and halocarbons are all heat-trapping gases, but water vapor is short-lived and mostly not anthropogenic, and the ozone layer is necessary to block out ultraviolet light. The 0.04% of CO_2 is the much publicized and worrisome number of **400 parts per million**[33] that we exceeded for the first time in 2013. It corresponds to about 3.2 trillion tons of CO_2 stuck in the atmosphere.

33. The total effective number for all greenhouse gases is probably 10% higher if one adds the methane and the NO_2.

What has happened between 2013 and 2018? According to the 2013-2014 Intergovernmental Panel on Climate Change (IPCC) report and the Global Carbon Budget Project, humankind had released approximately 36 billion tons of CO_2 to the atmosphere in 2013 from burning fossil fuels and cement production. Of this total, about 10 billion tons of CO_2 were absorbed by the oceans, leading to their increased acidification, and about 8 billion tons were absorbed by land sinks, mostly forests. The result was a net addition of 18 billion tons of CO_2 to the atmosphere in 2013. Given this unabated rate of annual accumulation, scientists predicted accelerated increases in global temperature and corresponding sea level rise, and storms, floods, droughts and wildfires with no end in sight. In response, the December 2015 Paris Climate Agreement, signed by 195 countries, pledged that the signatories would volunteer to adhere to a global plan of action limiting the worldwide average temperature increase to a maximum of well under 2 degrees Celsius (C), a daunting effort. Unfortunately, to complicate matters, on June 1st, 2017, President Trump decided to withdraw the U.S.A. from the Paris Agreement. The withdrawal will become fully effective in 2020 if it is not abrogated in the mean time.

In October 2018, the IPCC published an even more alarming special report. It stated that to limit the ultimate temperature increase to 1.5 degree C, the world would now have to decrease its green house gas emissions by 45% from 2010 by 2030 and 100% by 2050. In 2018 the global annual release of CO_2 had gone up to 37.1 billion tons from fossil fuels and cement production and 41 billion tons including changes in land use. The increases were attributed predominantly to China, the US, and India. The total atmospheric concentration of CO_2 was close to 410 parts per million. The average temperature increase of the earth exceeded 1 degree C above the pre-industrial age and the melting of polar ice caps was dangerously accelerating. Even a goal of a maximum increase of 2 degrees C was now in doubt. Coastal areas would be severely threatened by sea level rise, and tens of millions of people could be displaced.

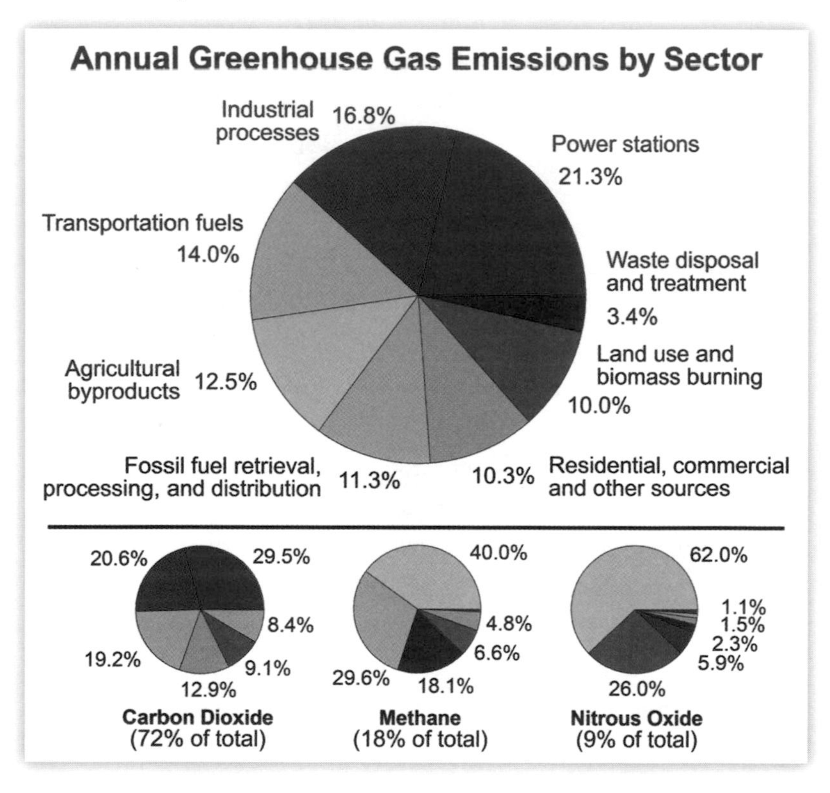

Fig.31 Annual Greenhouse Gas Emissions per Sector.

Technically, it is not that humanity doesn't have the means to save itself from impending disaster. It can in theory accelerate all the available remedies: energy conservation and efficiency, transition to cheaper renewable solar, wind and other energy sources, development of cheaper storage devices, electrification of all transportation, draconian reduction in the consumption of red meat, stopping deforestation, possible sequestration of green house gases, and as a long shot, geo-engineering means of reducing the solar energy received by the earth by seeding the upper atmosphere with light-reflecting particles. The question is: can all the countries of the world conjure up the political will and economic resources to act accordingly and to neutralize the powerful lobbies that are resisting such actions?

Some progress has been made at the December 2018 COP24 international climate meeting in Katowice, Poland: a common rule book has been signed to track, publish and reduce green house gas emissions by all participating countries, but there are still loose ends in how to fund these efforts and create market forces to insure these reductions.

The Social Sciences

Psychology

As a discipline, human psychology studies our minds, including conscious and unconscious experiences, and our behaviors. As part of these studies, it covers topics such as perception, cognition, love and sex, emotion, intelligence, attention, motivation, anticipation, brain functioning, learning, child development and rearing, interpersonal relationships, pathology, and therapies. Over time, psychology has branched out into various schools with different theoretical and therapeutic emphases on introspection, behavior, cognition, and humanism in general.

In what has preceded, we have already studied perception and cognition and how they inform our contact with reality. When it comes to love and sex, we are dealing with a complex mixture of nature and nurture. One of the few philosophers who had a very specific view of love between two mature adults was Arthur Schopenhauer (1788-1860). I am summarizing his view here because he may have come closest to a correct model. Schopenhauer asserted that when two people of the opposite sex feel attracted, they see in each other favorable combinations of traits that could result in a successful, good looking offspring. The attraction can lead to a one-night stand or a lifelong relationship, including marriage. Because the two separate human beings don't have infinite self-knowledge, they may not even be conscious of the mutual traits that attract them[34].

Either nature takes care of this attraction, possibly leading to the feeling of a romantic encounter, or marriages are arranged by families or friends. Dating services have to work by trial and error. Falling in love (note the idea of "falling" which implies inevitability) can happen at first sight or over a period of time but nature's not-so-secret goal is the same: procreation for the survival of the species. You will ask: "Why do people of the same sex also get attracted to each other, given that they can't produce a child?" Despite a huge amount of research, it looks like nobody really knows the origin of same sex attraction. A single gene is unlikely to be the cause because over generations, that gene would not be perpetuated since it won't lead to an offspring. Some theories invoke traumatic experiences in childhood but evidential support is insufficient. Other theories invoke hormonal anomalies in the mother's womb. Sigmund Freud and Alfred Kinsey thought we might all be bisexual but that for some reason, in the majority of us the heterosexual drive prevails. Or maybe Schopenhauer's model is incomplete. As Pope Francis has said: "Who am I to judge?"

The end-point, love, is probably the strongest feeling we can have, and it has no substitute. It not only produces children, it also

34. Given that bi-sexual reproduction dates back to about 1.2 BYA, it's had plenty of time to evolve to optimize its operation.

inspires novels, poetry, art, music, opera, and acts of bravery. When parents do have children, most of them love their progeny, and the children love their parents and seek their care and protection. Parental love, if expressed clearly and without ambiguity, is probably the greatest gift a child can receive. It can then translate into love of siblings, love of friends, love between partners or spouses, empathy, altruism, and love for humanity. It is our most empowering state of mind.

However, if we look at humanity, we see that parents are not perfect, that they don't always know how to act with their children, that their relationships don't always last, and that children don't necessarily get all the love they need. Hence, the ultimate results of parental love may not be as blissful as expected. Furthermore, as we will see below, the psychological and biological development of children may be fraught with some "built-in" obstacles that can cause them to develop mild or serious problems.

Independent of human love, our sex drive seems stronger than that of other species. While females of other species are *in estrus* (sexually receptive) during only specific annual intervals, women are receptive year-long, and so are men. Religions have tried to control and restrict sexual behavior for a number of reasons. Some restrictions arose to postpone pregnancy or out of hygiene considerations, others out of invented moral norms, male jealousy, punitive repression of eroticism, and so on. For example, as I write these lines, a group of Iranian young people are getting flogged (with 99 lashes each) for having participated in a co-ed dance party. Women are mistreated, subjected to "genital cutting" in many countries, and forced to cover their bodies to hide their attractiveness. Catholic priests are restricted to celibacy which may or may not be a factor in a number of them resorting to abusing children. The ultimate hypocrisy shows up when certain groups forbid the use of contraception and then want to punish women for contemplating an abortion. Unwanted pregnancy is a case where evolution, designed for survival, has gone awry: it's leading to overpopulation and potential disasters.

Coming back to our knowledge of the human brain in general, an amazing amount of progress is being made through studies using a variety of brain imaging scanning techniques. These go by abbreviated designations such as MRI, fMRI, MEG, PET, SPECT, and Ultrasound, which each use different methods to sort out our brain structures and functions. A current schematic is shown below.

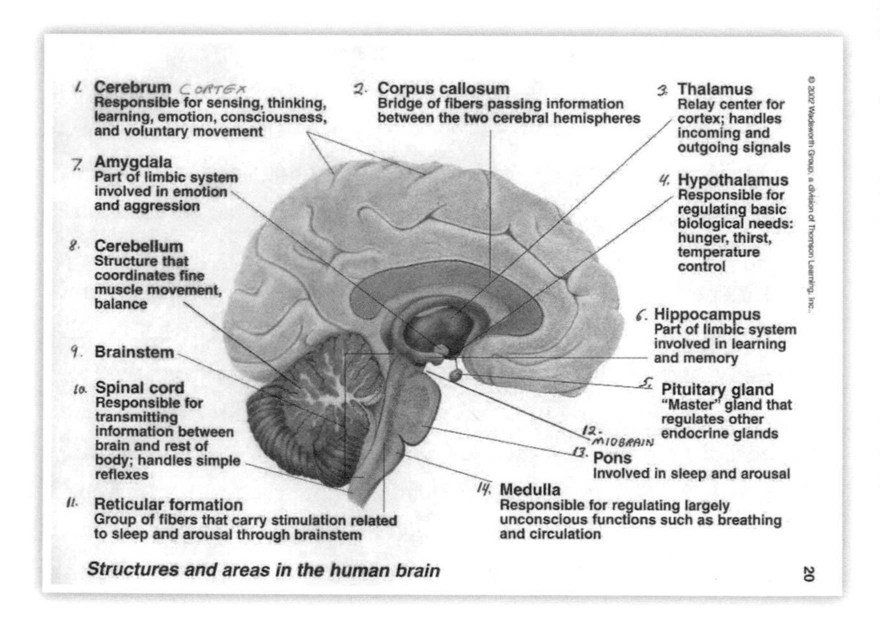

Fig.32 Current schematic of the human brain.

Some scientists like to identify our brains with computers (digital or analog). Even though brains perform certain operations similar to computers, I suspect the identification is wrong or certainly incomplete. Our brains are connected to our bodies, with which they constantly interact and produce conscious "qualia" as discussed earlier. No computer does that (at least for now). Animals, when observed, show limited repertoires but ours seems almost infinite and not always predictable.

Do we have free will? Here is a humorous answer:

When the Yiddish novelist Isaac Bashevis Singer received the Nobel Prize for Literature in 1978, he was interviewed by a series of reporters. The last one asked him:

"Mr. Singer, one final question, do you believe in free will?"

"Of course I do!"

"How come?"

"Do I have any choice?"

More seriously, I think the reason we have so much trouble with this question is that it suffers from semantic imprecision. In our lives, we constantly have to make decisions, large and small: where to live, what kind of work to do, who to marry, what story to write (if you are Mr. Singer), whether to be kind or nasty with somebody, what to wear, which movie to go to, which dish to choose from on a menu in a restaurant, and so on. In every one of these decisions, we feel we have some choice although sometimes, a keen observer may predict in advance what our choice might be. But that is not always the case. And it's not, as some scientists think (erroneously) because our neurons are subject to quantum mechanical probabilities. It's also not that in exercising our freedom, we will make a random decision. It's because we are able to think ahead and look at the consequences of our actions, and then base our decisions on weighing what those consequences might be. We are partially driven by instincts (also called algorithms these days) but we don't act by instinct alone, like animals predominantly do. Animals don't go on hunger strikes. Lions, dogs, and cats don't decide to become vegetarians. They don't write books. They don't know what it means to tell a truth or a lie, to act ethically or not, but we generally do. And that, it seems to me, is what counts when we contemplate this question. The more we can control our instincts and our obsessions, the freer we are.

What it also means is that the ability to make a considered choice is something we acquire with age. At birth, a child has only one primordial reflex: to survive. To do so it will cry when it is hungry or in distress. From then on, *nature, nurture, and intelligence* interact and gradually take over. Twenty long years or so are now needed for the child to become self-sufficient in modern society. As it grows up, it goes through a series of affective phases (see Freud below) and gradually develops physical, mental and social skills, and some degree of free will. Modern psychology has brought about enormous changes in child rearing guidance. A century ago, corporal punishment was still common. Dr. Benjamin Spock[35] (1903-1998) with his 1946 book on baby and child care changed all of this. But the fact that new "parenting" books are published every year, and that teaching methods are constantly being revised, means that we are still searching for better ways.

What happens when our brains malfunction? Some mental disorders like autism, Down syndrome, seizures, Parkinson's disease, Alzheimer's and other forms of dementia are due predominantly to biological factors. Others like anxieties, depression, bipolar disorders, paranoia, compulsions, phobias, eating disorders, anger, malign aggression, PTSD, alcoholism, and drug addiction are more likely traceable to developmental psychosocial or external factors. Schizophrenia can be due to either or both.

As history shows, diagnosing and categorizing mental diseases is a difficult enterprise. Starting in 1952, the American Psychiatric Association (APA) began to publish the first Diagnostic and Statistical Manual of Mental Disorders (DSM-I) listing 106 known mental disorders and "personality disturbances," which included ho-

35. See Dr. Benjamin Spock, "The Common Sense Book of Baby and Child Care," Duell, Sloan & Pearce, 1946..

mosexuality, and trying to diagnose and categorize them. Every few years, deletions and additions were made on the basis of changing criteria, and a new DSM would be published. In 1974, after strong protests by gay movements, homosexuality was removed from the list in the fifth printing of DSM-II and replaced by a new category of "sexual orientation disturbance." The same year, the APA decided that a new revision was needed to make it more consistent with an existing international classification adopted by the World Health Organization (WHO). This effort led to DSM-III published in 1980 which finally dropped homosexuality and eliminated *neuroses* from its list. By 1994, a new DSM-IV listed 297 mental disorders! And by 2013, the APA was publishing its fifth revision (DSM-V). Needless to say, the DSM "process" has traveled along an interesting but controversial road.

Psychotherapies

Today, professional help for patients with mental disorders is available from psychiatrists who specialize in administering drugs, and from psychotherapists who attempt to help them through verbal communication (talk). There are currently over one thousand different psychotherapy techniques "on the market," based on different models of the mind and possible approaches to introspection and behavior modification. Since it is impossible for me to explore all these techniques, I concentrate here on *psychoanalysis* because it is probably one of the deepest forms of psychotherapy, and the theories it is based on try to give a partial view of what I called *mental subjective reality* in Part 2. At the end of this chapter, I also briefly discuss the ideas of Alfred Adler and two other more recent therapies.

The founder of psychoanalysis, Sigmund Freud (1856-1939), devoted most of his professional life to studying the emotional problems of his patients by observing them, proposing models of the human psyche, and then coming up with a therapeutic method to help them. Freud over time modified several of his theories on account of his own investigations, observations, and, sometimes, the influence of some of his colleagues.

From about 1900 up to 1923, his central ideas were embodied in what he called the *topographic theory*[36].

This early theory assumed that our mental apparatus can be divided into three systems that differ from each other in their degree of accessibility to consciousness:

1) The *unconscious* which is driven by instincts and the *principle of instant pleasure gratification*; it is entirely dominant in early childhood, essentially inaccessible to consciousness, non-verbal and non-temporal.

2) The *pre-conscious* which is absent at birth, develops in childhood, acquires access to words and to consciousness, and is endowed with a censor that controls communication between the unconscious and the conscious; it can delay gratification subject to the *reality principle,* and is able, through the mechanism of *repression,* to block certain wishes of the unconscious in order to maintain mental health.

3) The *conscious,* which is aware of external stimuli and, via the mechanism of *attention,* thoughts, memories, and emotions brought by the pre-conscious; it also controls voluntary motor activity.

Gradually, Freud realized that this *topographic theory* was inadequate to deal with unconscious mental conflicts, neuroses and psychoses, and their therapy, and by 1923 he made a transition to a different model described by his *structural theory*. This model has some resemblances with his *topographic theory* but it works different-

36. For a detailed description of the topographic and structural theories and how they evolved in Freud's mind, see Jacob A. Arlow and Charles Brenner in "Psychoanalytic Concepts and the Structural Theory," International Universities Press, Inc., N.Y., N.Y., Second Printing 1973.

ly. It can be summarized as follows. The mind of every adult human being contains three separate internal "power centers" (also called "agencies") that form its structure: the *id*, the *ego*, and the *super-ego*.

1) The *id* is the seat of our instinctual drives and the source of our psychic energy; it comprises what we receive at birth and it acts autonomously and mostly unconsciously.

2) The *ego* is our sense of self; it develops gradually as we grow up, triggered by our perceptions, the physical activities we learn and perform, the acquisition of language, logical thinking, our controls and defense mechanisms, and our ability to test reality.

3) The *super-ego* is the seat of our internalized moral standards, commands and prohibitions that we acquire from our parents and our social environment. It can act both consciously and unconsciously.

These three "power centers" are not always at peace with each other and can create conflicts from childhood on. One of the first manifestations of these conflicts is felt by the child's *ego* through anxiety. This can happen when the *super-ego* unconsciously pushes back either on the *ego* or on drives from the *id*. Freud originally gave great importance to three early stages of the child's physical development during which its *id*, *ego*, and *super-ego* meet their first battlegrounds. He assumed that every child first goes from an initial oral stage at its mother's breast, followed by weaning, to an anal stage involving the control of his anal sphincter (toilet training), followed by a sexual/pleasure seeking stage involving its genitals. This third stage gives rise to what Freud called the Oedipus and Electra complexes. These are respectively the conflicts triggered by the fact that a boy is attracted to his mother and wishes to supplant his father, and a girl is attracted to her father and wishes to supplant her mother. These feelings, often unconsciously violent, may engender hostility and guilt in the child that, if not handled adequately, can leave it with long-term emotional traumas: specific feelings of deep loss at the end of the first stage, coercion and anger in the second, fear of punishment and castration in the third, and other more general feelings of fear, revenge, insecurity, inadequacy, severe narcissism, and difficulties of identifying with a parental model.

Fig.33a Sigmund Freud (1856-1939).

In Freud's model, the child may learn how to cope successfully with the above obstacles and reach adulthood emotionally unscathed, or along the way develop neuroses or psychoses. Neuroses result from an unsuccessful repression of certain instinctive drives, can involve anxiety and distress, and lead to certain compulsive behaviors. Psychoses, on the other hand, are much more serious because their patients lose touch with the outside world, which they perceive as collapsing. Psychoses are caused by an alteration of the ego function, incapable of dealing with painful conflicts with the *super-ego* and the *id*. To stave off anxiety, the patient regresses into a self-involved narcissistic state accompanied by delusions, hallucinations, nonentity and/or destructive anger.

Fig.33b Freud's famous couch where he placed his patients, and the arm chair on the left where he sat during psychoanalytic sessions.

To treat his patients' neuroses or psychoses, Freud prompted them to relax on his famous couch and talk to him as freely and spontaneously as possible. In his analytic process, he used two parallel tracts: 1) discover and bring to the surface the likely *defenses* raised by the patient's *ego* to deny its repressed guilt and aggressive feelings, and 2) facilitate the patient's access to his/her unconscious memories via free language associations, occasional inadvertent parapraxes (so-called Freudian "slips"), and trying to remember and interpret dreams. Dreams, according to Freud[37], have a *latent*

37. Freud's seminal "Interpretation of Dreams" was published in November 1899 and reappeared many years later in his more readable popularized version titled "On Dreams" with a biographical introduction by Peter Gay, W.W. Norton, 1989. His insights into dream formation were mostly interpreted as wish fulfillments and probably not complete. Dreams, "the guardians of sleep" may sometimes serve purposes other than wish fulfillments. For example, certain dreams may be created simply to make an external happening (e.g., a telephone ring) understandable while asleep. Also, Freud did not yet know that dreams occur predominantly

content which contains unconscious information about childhood traumas and wish fulfillment, and a *manifest* content that we construct through "dream work" under *ego* and *super-ego* censorship and experience consciously as a story or a scene while we sleep. The psychoanalytic process may frequently lead the patient to *regress* into an earlier stage of its *ego* development and is supposed to be facilitated by *transference*, the presumed mechanism by which the patient transfers his or her *super-ego* to the empathetic analyst. The latter can then interpret what the patient says and bring his or her traumas back to consciousness, thereby facilitating both understanding and relief from pathological and compulsive behavior.

What is the track record of this therapy? Psychoanalysis was immensely popular for some time even though it is slow, expensive, unpredictable and chaotic, and depends considerably on the skills of the analyst. Freud must have had an uncanny imagination to construct this entire edifice but the theory and practice of psychoanalysis illustrate how difficult it is to apply the scientific method (observations, model making, and verification through experiments) to mental health. How can one check out and measure the tortuous interactions between the *ego*, the *id*, and the *super-ego* that Freud postulated? Or verify the mechanisms of memory repression into our "hermetic" unconscious? Why did evolution build into our minds these obstacles to our mental health? On the other hand, why wouldn't "talking" to a sympathetic analyst for several years bring some relief to internal conflicts? It lets us peek into parts of our *mental subjective reality*, even if it doesn't guarantee a complete "rewiring" of our psyche. Psychoanalysis has progressively evolved since Freud's death and its current practitioners continue to generate new ideas and techniques. Even if the therapy is no longer as fashionable as fifty years ago, it still has a very positive track record

---

during rapid eye movement (REM) sleep when our bodies lose most muscle tone (REM atonia) and our brains reset our memories and prepare for the next day.

and produces beneficial results, particularly for people who have the time, money, and motivation to practice deep introspection.

Looking ahead, what will be Freud's legacy? During his career, he had many collaborators, admirers, and disciples over whom he tended to rule autocratically. Some of Freud's claims of "cures" have been disputed and he has been accused of fudging some of his data. Over the years, he has also had many critics and detractors[38]. It seems to me that however one views Freud, warts and all, his contributions to humanity will long be remembered. By pointing out the awakening of child sexuality at a very early age, he exposed a taboo that permeates all societies. His ideas on the unconscious and his interpretation of dreams will not go away. Beyond psychoanalysis, Freud made it acceptable to talk openly about sexuality and its implications. He also pointed out that many religions are based on the belief in a strong father figure that must be worshipped and feared. Repressive leaders can exploit this fear to attain and retain power. If you have any doubts, just look at the insane religious feuds in today's world.

One of Freud's early collaborators and colleagues in the psychoanalytic movement was Alfred Adler (1870-1937), also an Austrian, born in the outskirts of Vienna. Freud and Adler shared ideas for many years and respected each other but eventually parted ways in 1911 after some disagreements.[39] Adler then in 1912 founded the Society for Individual Psychology. In so doing, he did not abandon Freud's ideas on depth psychology, the unconscious and the importance of dreams in therapy. However, he gave new emphasis to

38. One of the most ferocious ones is Frederick Crews who recently published a book titled "Freud: the Making of an Illusion," Holt, Henry and Co., Inc, 2017.

39. In retrospect it is perhaps unfortunate that Freud and Adler parted ways because their theories and therapies were not as incompatible as they might have appeared at the time. As mentioned here, Adler had many ideas that could have enriched and complemented Freud's views and practices.

social interactions beyond sexuality and came up with the concept of the ***inferiority complex***. This feeling of inferiority can develop in early childhood because of the sheer difference in size and physical strength of the child and its parents, or because of parental mistreatment or ignorance of the disproportionate power relationship. Adler believed that treating patients suffering from this complex and its consequences required discovering and interpreting their early memories, including their interactions with competing siblings. He not only thought that these techniques could be useful to treat patients *a posteriori* but also preventively by training potential parents and lay people before they had children. He was also one of the first psychologists to advocate in favor of feminism and a more balanced relationship between men and women. Adler was greatly admired by other psychologists like Rollo May, Abraham Maslow, Karen Horney and Erich Fromm.

Before moving on, I want to mention two other current psychotherapies also relying on talk that are popular and shown to be effective in treating some, if not all, problems. One is *Psychodynamic Psychotherapy* which is based on most of Freud's theories, and dates back to Carl Jung, Alfred Adler, Otto Rank, and Melanie Klein. It demands a very strong bond and degree of trust between analyst and patient, but is limited to a shorter treatment time, maybe one or two hours per week for a year or less. Its goals, like psychoanalysis, are the alleviation of acute symptoms, the promotion of internal growth, a sense of freedom from compulsions, and improvement in personal relationships. The other one is *Cognitive Behavioral Therapy (CBT)*, based on the belief that psychological disorders are the result of maladaptive thoughts and behaviors ingrained in the patient also starting in childhood, but not necessarily buried in the unconscious. The role of the therapist is to identify these maladaptive factors and to modify the patient's behavior by providing him or her with new thoughts, skills, and coping mechanisms. It does not assume free will. CBT has evolved through several "waves" or phases differing in how they view and handle symptoms. It appears to have a good track record in treating depression and anxiety but its effect on other disorders is more controversial.

To summarize, the goal of psychotherapies is to help us function rationally. It doesn't "cure" us like an antibiotic cures us from an infection but it can free us from various forms of destructive and compulsive behavior. It doesn't automatically offer us "happiness" but it can open the door to a happier life.

Thoughts on Power, Competition, Sports and Patriotism

Whether or not we all grew up with an inferiority complex, it is most likely that we have all felt inferior physically, mentally or emotionally at some point in our lives. Were these feelings destructive or beneficial? It obviously depends. If they were paralyzing, led to discouragement, jealousy, lust for power or violence, obviously they were destructive. On the other hand, if they motivated us to become more resilient, stronger, more moral and achievement-oriented human beings, they were beneficial.

From their earliest age, children want to be supermen and superwomen. Exercising and participating in competitive sports to improve our bodies and becoming stronger, as long as they don't become obsessions, are surely constructive activities. Competing with our siblings for the love and attention of our parents is normal but when it becomes extreme, it can be a symptom of dysfunction. Studying hard in school in competition with other students is not necessarily bad if in the process we become more proficient intellectually or skillful. Wanting to be "first in class" may set a good example but can become pathological if it stems from a need to show one's superiority to everybody else. Developing leadership skills can be positive and inspiring, but as we will see in a later chapter, learning how to cooperate with others may be even more important for society. Competing in business drives the market economy. Attaining power may be an antidote to feeling inferior but as we know, uncontrolled power corrupts.

Another antidote to feeling weak and inferior is the sense of belonging. We are all familiar with the satisfaction and comfort of belonging to a family, a group, an institution, a business, a religious congregation, a political or military organization, a club, a sports team, a city, a region, a state, or a nation. It makes us feel protected, stronger and can provide us with a meaning we may not have in isolation. We also know that these feelings can be double-edged, depending on circumstances. Rooting for a team, whether at the local, national or international level, is fun and gives us a sense of satisfaction and power when we win but frustration when we lose. Just think about the so-called World Series for baseball and the Annual Super Bowl for football in the U.S., the Soccer World Cup and the Olympic Games, worldwide. They trigger admiration for skill, a tremendous sense of competition, the glory of victory, and occasional mindless violence!

Patriotism is another perfect example of what it means to belong. It can bring the best out of citizens of a nation and the worst through the abuses of nationalism. And without "esprit de corps," military training, organization and sacrifice in the name of a cause would be impossible!

Paleoanthropology, Anthropology, and Sociology

Dr. Peggy Golde[40], a long-time anthropologist friend, used to joke, "Anthropology is where they go by boat, sociology is where they go by bus." I would add that paleoanthropology is "where they go by jeep or by foot."

More seriously, paleoanthropology is the study of the origins and reconstruction of evolutionary kinship lines of hominids. Anthropology is the study of past and present social and cultural norms of human groups, tribes, and populations often, but not always, living in relative isolation from the modern world. Understanding them requires an open mind devoid of *a priori* theoretical models. In the U.S., anthropology also includes archaeology, which in Europe is considered a separate discipline.

Modern sociology was launched by the French philosopher Auguste Comte (1798-1857) in the 19th century. Comte was followed by several famous pioneers such as Herbert Spencer (1820-1903), Emile Durkheim (1858-1917), and Max Weber (1864-1920). Nowadays, sociology typically tries to study the socio-dynamics of authority and power structures in society, economic classes, ethnic and race relations, education, family and gender roles, special interest and professional groups, and networks and institutions. Admittedly, the boundaries between social anthropology and sociology are somewhat fuzzy. Sociology as a science is still a work in progress. Sociologists have made great strides in observing phenomena in the above categories but they are still searching for the correct structures and models to explain and predict their observations. Nor are they very successful at prescribing remedies to our social woes. Quantitative measurements assisted by computer and statistical techniques will no doubt bring significant advances

40. See "Women in the Field: Anthropological Experiences" edited by Peggy Golde, University of California Press, 1986.

to sociology in the future. A considerable amount of data is also currently being collected by the providers of social media, who no doubt exploit them for commercial and political use.

Paleoanthropology of our ancestors is one of the most fascinating detective stories in progress in the social sciences. As "forensic" evidence, it uses fossils (petrified skeletal remains), bone fragments, teeth, footprints, stone tools, artifacts, settlement localities, and, more recently DNA, when available. The evolution from our common ape ancestors to *homo sapiens* roughly 5.5 million years ago (MYA) that took place in Africa was already sketched in the introduction. It started with the Australopithecines (from the Greek for "southern apes"), a "hominid" genus appearing around 4.5 MYA in Eastern Africa and eventually spreading throughout the continent, until it became extinct around 2 MYA. During this time a number of Australopithecine species emerged, of which two of the most famous specimens are the A. Ramidus "Ardi" (roughly 4.2 MYA) and the A. Afarensis "Lucy" (roughly 3.2 MYA). The exact evolutionary lineages of probably several parallel contemporary hominid species have not yet been exactly sorted out, but the genus *homo* almost surely is related to one of them. With the gradual development of tool-making, *homo habilis/ergaster* took over around 2.5 MYA. This date is defined as the beginning of the Paleolithic era or Stone Age. *Homo habilis* was followed by the bipedal *homo erectus* (around 1.9 MYA). *Homo erectus* was probably the first to learn how to control fire. This discovery over millennia had several important effects on their users: it lengthened the duration of their days by enabling social groups to gather around hearths after dark (something other animals couldn't do), and it eventually changed their diets by allowing them to cook (soften) roots and meats. The effect of this was that their intestines became shorter and that their brains possibly grew even faster. Around this time, some of these hominids began to leave Africa and settle in northern China and western Indonesia, and subsequently in Europe as far as England. Another group called *Homo heidelbergensis* left Africa for Europe around 500,000 YA and evolved into the Neanderthals around 200,000 YA. All these groups are extinct now. Meanwhile, *Homo sapiens* is believed to have evolved in eastern Africa, between 300,000 and 200,000 YA. The entire sequence is illustrated in the diagrammatic time-lines on the next page[41].

Specific dates are still changing. Even the sequential ancestries of the more recent overlapping species are still controversial and not well established. For example, it is not certain that *homo erectus* is the descendant of *homo habilis,* and the onset of bipedalism is still not clear.

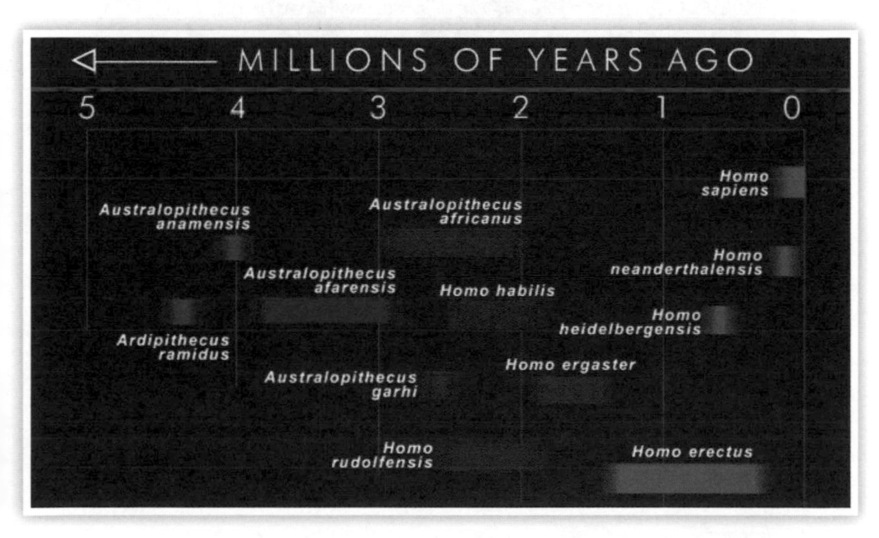

Fig.34 Time lines of human evolution starting from 5 million years ago to today.

41. For more information see for example Richard E. Leakey and Roger Lewin's "People of the Lake: Mankind and its Beginnings," Avon Books, 1979, or watch the fascinating NOVA movie "The Dawn of Humanity."

Homo sapiens very rapidly developed modern bone artifacts, knife blades, projectiles for hunting and fishing, piercing tools, drilling tools, and engraving tools. These new technologies increased their users' resources and caused their populations to grow significantly. Starting around 70,000 YA, they began to move out of Africa in droves and eventually spread over the entire planet[42].

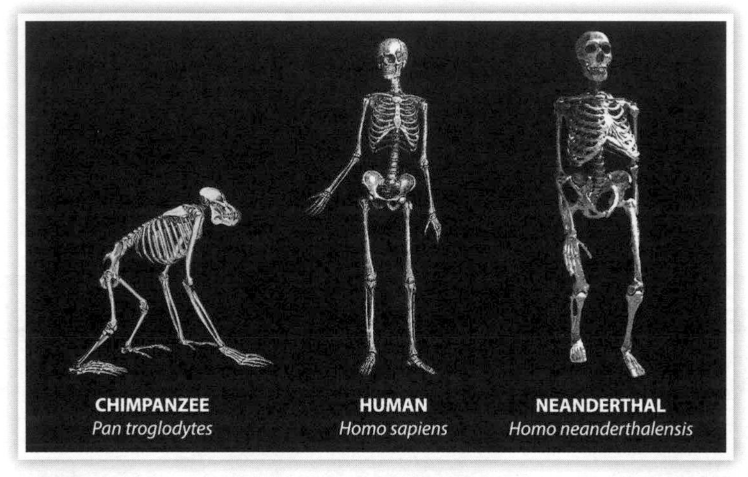

Fig.35 Comparative skeletal structures of chimpanzee, homo sapiens and homo neanderthalensis (extinct).

Meanwhile, our remote cousins, the Neanderthals, had adapted to the Ice Age in Europe. They lived in relatively small groups extending from Israel through Germany to Spain. Recent findings show that they were considerably more advanced than originally thought, had larger heads and stronger bodies than us, used fire, buried their dead, and had 99.7% of their DNA identical to modern humans. It is assumed by some researchers that they interbred with humans, but this is contradicted by others. There are many hypoth-

42. See a recent account of how the various Homo sapiens migrations to Europe, Asia, Australia and America are being tracked by decoding our early history with DNA in "Profiles in Science", Eske Willerslev, New York Times, May 17th, 2016.

eses as to why they became completely extinct around 25,000 YA: climate change, starvation, disease, cannibalism, violent competition with far more numerous *Homo sapiens,* inability to speak, or maybe all of the above.

Beginning around 40,000 YA (38,000 BCE), it appears that wherever *Homo sapiens* moved between Spain and Indonesia, they came up with one similar idea: cave paintings. What these paintings show is that these people must have had very similar perceptive systems and *representative reality* to ours. Whatever views they had of the world they lived in, their motivations to create these paintings seem similar. It is not known whether the artists were male or female, young or old, but they must have studied and trained to commit their images (mostly animals) to memory ahead of time. Indeed, it appears that these animals were not dragged alive into the caves as models. The artists also didn't leave any indications that they themselves lived in most of the caves (although some caves may have been dwellings). They used similar pigments, including red and yellow ochre, hematite, manganese oxide, and charcoal, sometimes incising the silhouettes of their subjects in the rocks, or engraving them.

Given the huge geographical distance between Spain and Indonesia, we can't assume that there was communication between the artists at Altamira in Cantabria and Maros in Sulawesi Island. However, since they were all hunter-gatherers, their common interest in animals and hunting is not surprising. The drive for survival and need to find food must have been at the top of their *mental subjective reality* and daily preoccupation. It is tempting also to believe, as some anthropologists do, that some of these artists had metaphysical anxieties which drove them to "animism." Animism is the belief that humans, animals, objects, and even phenomena like thunder and wind have souls and kinship relationships. The artists may have wanted to venerate these animals graphically.

In the absence of written records or explanations, we may never know for sure.

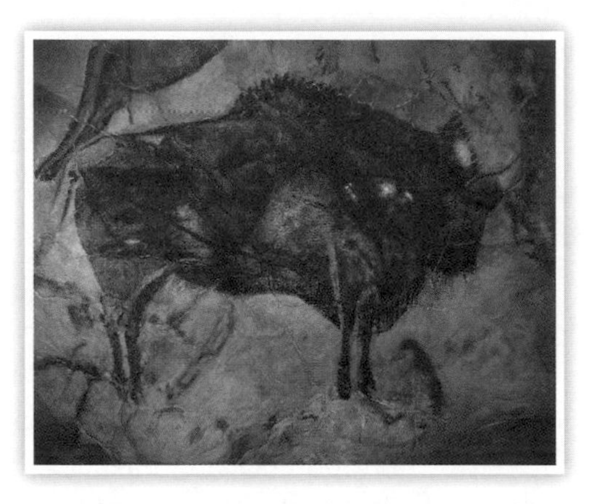

Fig.36 Cave painting of bison at Altamira, Spain.

Fig.37 Hall of bulls at Lascaux's cave, France.

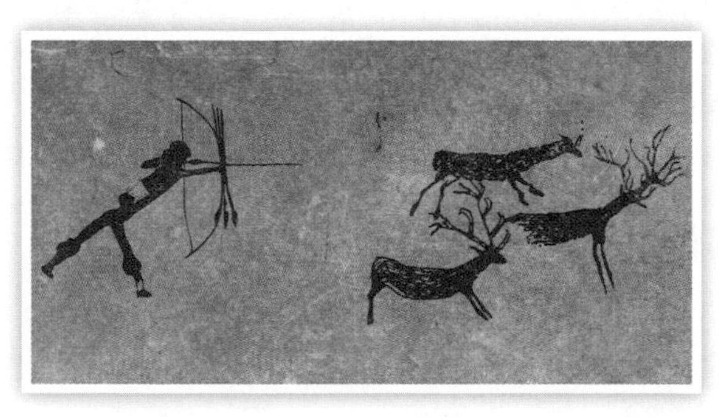

Fig.38 Paleolithic cave painting of hunting scene.

Skipping over the poorly defined Mesolithic era, the Neolithic era began around 10,000 BCE. In broad terms, what defined it was the revolutionary advent of agriculture and domestication of animals. These two factors completely changed the way of life of humans, allowing them to become sedentary, first in small and later in larger groups, and eventually in cities and empires. Dwellings and pottery gradually made their appearance. Mesopotamia was the earliest site of these developments that inspired other major inventions such as cereal crops, the wheel, cursive script, mathematics, and astronomy. New technologies also brought along labor specialization, slavery, professional soldiers, absolute leaders, and more stratified social organizations that until then had been more egalitarian.

The Bronze Age (bronze being a copper-tin alloy) started around 3200 BCE in the Aegean, Mesopotamia, and Egypt where it was accompanied by early forms of writing systems (cuneiform and hieroglyphics, respectively). The Bronze Age spread to the Levant, India, China, Southeast Asia, Russia, and Europe as far as Great Britain and Scandinavia. The technology even appeared independently in South America and Mexico. Not too long thereafter, the discovery and occasional use of iron also appeared in Anatolia and India. It is interesting to note that the next age, the Iron Age, actually took off around 1200 BCE when tin became scarce for some time, and smelted iron was able to replace bronze to make all kinds of artifacts and weapons.

The time boundary between what we call Prehistory and History is fuzzy because the demarcation that is used is the appearance of written records, which didn't happen simultaneously in all parts of the world. However, it is generally agreed that the Sumerian/Akkadian civilization marks the beginning of History during the Bronze Age in Mesopotamia or Arabia, and that by the advent of the Iron Age, written records had sprung up everywhere, putting an end to Prehistory. Interestingly, many of the early records had to do with properties and taxes! From here on, we have to look at Anthropology and History for the study and reality of human behaviors and events. In what follows, I will concentrate on two issues: the advent of war and the advent of religion, both of which have been pervasive throughout History.

What is War?

War is armed conflict between groups of humans within a geographical area, inside a nation, or between two or more nations. It can also manifest itself in the form of rebellions, revolutions, civil wars, colonial wars, wars of independence, military interventions by external forces inside a sovereign nation, genocide, and terrorism. War is a horrible collective human activity that gives license to violence, killing, and other destructive behavior that would normally be considered criminal within a society at peace. During the 20th century alone, wars between two or more nations have killed about 95 million soldiers and civilians, civil wars probably about 57 million people, and various genocides like in Armenia and the Holocaust at least 10 million.

Anthropologists struggle with the question of whether war is an innate or learned, cultural form of behavior. Margaret Mead (1901-1978) thought that if one could find even a single group of people like the Arapesh in New Guinea that are generally non-violent and don't engage in armed conflict, then war cannot be an innate form of behavior, and must be learned. Psychologist and sociologist Erich Fromm (1900-1980) further pointed out in his extensive study of *The Anatomy of Human Destructiveness* (1973) that one must be careful to distinguish between various forms of aggressiveness. *Benign aggression* is biologically adaptive as a response to a threat and is designed to remove it by either fight or flight. It is common to all animals and humans. *Malign aggression, i.e.,* destructiveness and cruelty, is not innate and is the result of frustration and pathologically punitive upbringing.

Fig.39 American anthropologist Margaret Mead (1901-1978).

In the last few decades, some anthropologists have questioned Margaret Mead's conclusion on the basis of various comparisons. Those who have studied chimpanzees (our cousins from 5.5 MYA) have determined that while they don't carry any kind of weapon, they do tend to not be particularly cooperative with each other and form rival territorial groups. The sizes of the territories they guard are a function of the density of available food and prey necessary to sustain their members: *i.e.,* if food and prey are scarce, the territories are larger. Occasionally, members of one group attack and murder members of a group from a nearby territory. But it's not mass murder and the attack is short-lived. Similar territorial behavior is observed between packs of wolves. Otherwise, most other carnivore animals kill their prey without hesitation or pity but don't kill members of their own species, even if they belong to a separate group. The bonobo apes, for example, which are not genetically distant from the chimpanzees, don't engage in group violence.

There are conjectures, but no records, about whether homo habilis and homo erectus practiced warfare. They no doubt must have had conflicts; the males must have fought over females, there were probably occasional fratricides (Cain and Abel type), and fights over food. But they also must have felt very fragile and vulnerable to animal predators, and spent much of their time running away from them rather than defending any turf and attacking each other's groups. The cave paintings from 40,000 YA depict hunting scenes with weapons but no fights between humans. However, a recent discovery[43] at Nataruk, a rich and luscious area west of Lake Turkana in Kenya, seems to indicate that at the end of the Paleolithic era, around 10,000 BCE, a band of 27 hunter-gatherers were brutally murdered with bows, arrows and clubs in a raid by another unknown group, possibly competing for resources.

As discussed earlier, agriculture and the Bronze Age brought along sedentary life, new social organizations, economic resources, and technologies. Ever since, conflicts between groups, nations, and empires over **space, resources, populations, and ideologies** have had the potential of leading to war. But not all conflicts necessarily lead to war. Still today, if the leaders of these respective groups are wise and not overly ambitious, they are able to resolve their conflicts peacefully. In his non-violent movement to liberate India from British control, Mahatma Gandhi (1869-1948) gave an admirable example to humanity.

43. See M. Mirazon Lahr et al. in "Inter-group violence among early Holocene hunter-gatherers of West Turkana, Kenya" in Nature 529 394-398, 21 January 2016.

Fig.40 Mahatma Gandhi (1869-1948).

If leaders are mediocre or pathologically aggressive like Hitler, they will blunder and/or try to appeal to their subjects by whipping up false patriotism through demagoguery and fear, and lead them to war[44]. Invariably, they will claim that they are responding to a threat and will never admit to being the aggressor.

The instruments and military strategies of warfare have evolved immensely with time. Rocks, bows and arrows, spears, daggers, lances, swords, chariots, infantry, cavalry, huge standing armies, muskets, rifles, machine guns, cannons, tanks, navy and aircraft, and chemical, biological, and nuclear weapons; all these weapons have been used. The advent of sovereign nation-states, which emerged from a need to govern and organize the lives of people in a coherent way, has also given rise to nationalism. Unfortunately, nationalism has made it easier for a leader to start a war, even in a democracy. I

44. See John G. Stoessinger, "Why Nations go to War," 11th Edition, Wadsworth (2011).

personally think that ascribing some "war gene" as a cause of such a complex phenomenon does not get us very far. Evolution certainly didn't give us a gene to build a nuclear bomb! Wars today may start UNLESS there is a factor that deters the leaders from them, such as fear of defeat or some kind of international law or institutional restraint. I will come back to this point later.

What is Religion?

Religion is not easy to define: there are currently about 10,000 different religions in the world and many of them have evolved over time. The five major religions are Hinduism, Buddhism, Judaism, Christianity, and Islam, and around 75% of the world's population observes at least one of them. Each of these religions started with an original teacher, prophet, and/or messiah. All three monotheistic religions, Judaism, Christianity and Islam recognize Abraham as their earliest common patriarch and founder. Adherents to a given religion tend to share a common system of beliefs, rituals, and codes of behavior. The beliefs generally, but not uniformly, include the existence of one or several super-natural entities, the commitment to a faith, and an explanation of the origins of the universe and humanity. The rituals include places and forms of worship, sacred scriptures, and veneration of the super-natural entity or entities. The codes of behavior prescribe norms of personal and social ethics, prayers and/or meditations, and details about human relations, marriages, attires, and possibly diets. Buddhism stands apart in that it is not based on worship, but enlightenment through meditation and ethical behavior. It believes that human suffering is caused by desires and cravings which can never be fully satisfied, and that suffering can be relieved by practicing introspection, asceticism, and striving ultimately for the attainment of *nirvana,* a state of total nothingness and bliss.

Anthropologists, sociologists, philosophers, and historians have all tried to explain how religions have come about and the functions they perform. My personal view can be summarized in five points:

1. Religions appeal to humans because they offer them an explanation for which they yearn, namely where they and the universe come from, and how their lives fit into it.
2. The fear of death is terrifying and the hope that there may be some afterlife or some form of survival for our souls is a reassuring idea onto which to latch.
3. Religions give humans some solace when life is disappointing, frightening or cruel, and reassure them that a compassionate God or other deities are watching over their destinies and protecting their lives.
4. Religions require their adherents to accept a code of behavior often dictated by a charismatic authority figure (on earth or in heaven) with rules that they may not otherwise be inclined to obey.
5. Religions create a sense of community and power among their adherents that can be exploited for good or for evil.

Let me consider these points, one by one.

The first point cannot be proved or disproved by science. Astrophysics currently has shown that our universe started with a Big Bang roughly 12.8 billion years ago, but nobody knows what happened before the Big Bang or whether there exists an infinity of universes. This point is not about to be elucidated any time soon, and if people feel better in their *mental reality* that some deity created the universe, I can understand. As for the advent of life on earth, it should be no secret that I believe it was the beautiful work of evolution. Some scientists like to invoke the so-called "anthropic principle" (the word "principle" is actually a misnomer), an observation that shows that the most minimal changes in some of the fundamental constants of physics would have made life as we know it impossible on earth. This may be true but those are the fundamental constants we have and ours is the life that has ensued, without the need for intelligent design.

The second point, *i.e.* the fear of death, will always be with us. I personally do not believe in any afterlife but I have no better comfort to offer to those who want to believe in it, except maybe to do something of lasting and transcending value on earth or to practice Buddhist meditation.

The third point, that life is rough and can be horribly unfair, is obvious. The question often comes up among religious people: "How can a caring and compassionate God allow so much suffering and injustice?" and a frequent answer is that "God is testing us." I simply don't see how this theory can be correct: was the Holocaust a test?

The fourth point is a practical one. It seems to me that humans should behave themselves in an ethical way, whether they believe in God or not, because that's the only way life in society can be tolerable. The argument that the fear of God's punishment "tames the beast in us" may of course work sometimes. A case in point is Moses and the Ten Commandments (note the unmistakable word "Commandments"). This story first appeared in the Old Testament which was written by scribes in Babylon when the Israelites lived there in exile, probably hundreds of years after their conjectured exodus from Egypt. I say "conjectured" because Israeli archeologists today have found absolutely no record of Israelite tribes ever living in Egypt in that period, or that Moses ever existed. But you must admire him for condemning his fellow tribesmen for worshipping a golden calf while he was meeting with God. Clearly, he already foresaw the sins of Wall Street! Having said this, it is unfortunate that one has to appeal to fear to make humans behave. Besides, criminals get away with murder and there is no evidence that God punishes them on earth. In heaven, of course, I don't know, but then there is always hell.

The fifth point presents us with a dilemma. There is no doubt that many religious people behave themselves ethically and are loving, generous, tolerant, and kind with their fellow human beings. Pope Francis, within some institutional constraints of the Catholic Church, is definitely a great force for good. On the other hand, it is also obvious from past and present experience that horrific crimes have been and are being committed in the name of religion. So how do we deal with this dichotomy? As it turns out, in countries where religions have been banned, *e.g.*, under Communist regimes, equally heinous crimes have been committed.

It seems to me that the only viable conclusion to be drawn from this discussion is that people should be free to practice any religion they want as long as they are tolerant and do not try to impose it by force on believers of other religions or non-believers. They must also respect the strict separation of church and state, secular education, the rights of minorities, the equality of men and women, and the right to contraception and abortion. This latter issue is one where both religion and evolution have gone wrong. If the world population reaches 10 billion people, life on the planet will become hard to sustain.

Fig.41 Michelangelo's "Creation of Adam" in the Sistine Chapel ceiling, Rome, Italy.

Fig.42 Siege of Antioch (1097-1098) by the First Crusade of Christians against Muslims. Antioch was a fortress in Syria occupied by the Muslims trying to bar the Crusaders from reaching Jerusalem.

Economics

One of the prevalent definitions of economics is that it is the study of the factors that determine the production, distribution and consumption of goods and services. I think this definition is incomplete without specifying what these factors are, namely the respective values of labor, land, commodities, technology, and capital. Economics is studied at two levels: *microeconomics, i.e.,* what happens at the level of the individual, the firm, and its market, and *macroeconomics, i.e.*, at the level of the economy as a whole (domestic or international). Furthermore, unlike the physical sciences, economics is normative in the sense that it is used to answer "not just what is, but also what should be." And "what should be" is in the eyes of the beholder: *e.g.*, social justice, or efficiency, or profits, or growth, or power, or other goals.

While modern economics creates models and tries to verify them, it is still far from an exact science because a) selecting and measuring the relevant functions and variables is not always obvious, b) alternative tests are hard to perform, and c) the "economic landscape" is constantly changing with geographical location, time, and innovation. Given these uncertainties, politicians can easily latch on to different interpretations and opposite remedies to support their ideologies.

It was Thomas Aquinas (1225-1274), famous Dominican friar and influential theologian, who surprisingly contributed to early economic thought by introducing the concept of a "just price." He thought that the price of an object should cover its cost of production, but that it would be immoral for a seller to raise the price of the object just because the buyer was in dire need of it. Between the 16th and 18th century, there were two preponderant schools of economics: *mercantilists* who believed that the wealth of a nation depended on how much gold and silver it could accumulate through mining, trade and protective tariffs, and *physiocrats* who thought that only agriculture could create an economic surplus and growth. The latter recommended a policy of laissez-faire and a single tax on land owners.

Modern economics began with Adam Smith (1723-1790), moral philosopher and a key figure of the Scottish Enlightenment. His *Theory of Moral Sentiments* (1759) and *Inquiry into the Nature and Causes of the Wealth of Nations* (1776) are still considered among the most influential works in the field of economics today. In his first book, Smith praised the constructive aspects of "mutual sympathy and trust" that develop between human beings when they deal with each other, akin to what today is called "empathy." This idealistic view of human relations clashed somewhat with his belief, expressed in the *Wealth of Nations,* that individual self-interest is what runs the engine of the free market economy. The result is a system driven by a self-regulating "*invisible hand.*" Smith believed that if resource owners and managers of *specialized* labor, land, and capital were unleashed in a truly *competitive* system, prices of goods and services would drop to their lowest possible levels, providing the greatest benefit to the largest number of consumers. He warned about the negative effects of monopolies, and thought that the state was necessary to provide public goods, education and infrastructure, enforce laws and contracts, regulate banks, and raise funds for defense. Smith also believed that free trade between nations would be of greatest mutual benefit and that restrictive tariffs would only hurt them.

Fig.43 Adam Smith (1723-1790), philosopher of the Scottish Enlightenment, author of the "Theory of Moral Sentiments" and the "Inquiry into the Nature and Causes of the Wealth of Nations."

Adam Smith's most illustrious successors straddling the 19th century were Malthus, Ricardo, Mill, and Marx.

Thomas Robert Malthus (1766-1834) in his pessimistic book, "An Essay on the Principle of Population" (1798-1826), argued that world population (then 1 billion) would grow geometrically while the food supply would grow arithmetically. This would lead to the impossibility of providing adequate food subsistence to humanity in the long run and result in what became known as the Malthusian Catastrophe. His prediction did not materialize because he underestimated the capability of agriculture to grow as fast as it did. Unfortunately, it is still possible that history will prove him right two or three centuries later when world population may reach 10 billion and sustainability will no longer be limited just by food.

David Ricardo (1772-1823) started his career on the London Stock Exchange where he amassed a large fortune. After reading Adam Smith's *Wealth of Nations,* he became fascinated by economics, to which he made remarkable contributions for the rest of his life. The first one was his work on *The High Price of Bullion* in 1810 in which he pointed out that the Bank of England, by being unrestricted by law, was circulating far too many banknotes compared to how much gold it held, thereby causing inflation. The policy of adjusting the quantity of money in circulation that he analyzed is what is called *monetarism* today. Around 1815, Ricardo came up with his essay favoring free trade against the *Corn Laws*. He showed that these laws, which raised tariffs on the import of wheat, only benefitted the land owners but not the consumers. In 1817, he published his magnum opus, *Principles of Political Economy and Taxation*, in which he analyzed the laws governing what could be produced by landlords, workers and owners of capital. On all these topics he had frequent discussions with Malthus, but didn't generally agree with him. One of the questions that remained open after Ricardo still lingers today: "Is the market economy just efficient at optimizing the allocation of resources but blind to the distribution of income?"

Years later, John Stuart Mill (1810-1873), the precocious son of philosopher James Mill and admirer of utilitarian Jeremy Bentham, became involved in this debate. Ethically, he strongly believed in liberty, non-conformity, the emancipation of slaves, and women's equality and right to vote. On the other hand, very strangely, he was not against despotism when it came to Westerners ruling "*barbarians*," as he labeled the people of India and China. In economics, he viewed capital as the product of "accumulated former labor," which was necessary for economic growth and wealth, and could be used in part to increase current wages. Mill was wary of the potential tyranny of the state but also thought that the state had to intervene to improve income distribution. In some ways, he was an early socialist, as he believed in labor unions and farm cooperatives, and he was also an early environmentalist.

The biggest upheaval in the economic debate of the 19th century broke out with Karl Marx (1818-1883). After the American and French Revolutions and the Industrial Revolution, the aristocracy as a class had been replaced by the bourgeoisie in parts of Western Europe and America where capitalism flourished. But capitalism did not solve the problem of poverty, and was often seen to increase it. The working classes of some industries, including child laborers, lived in squalor. Stemming from a bourgeois family himself, Marx argued that the workers were paid only minimal subsistence wages, far below what they really contributed, and that the ruling classes exploited them by pocketing the difference, the *surplus earnings*, for their own profit. He became convinced that history is nothing but a sequence of economic class struggles: as feudal lords had been succeeded by the aristocracy, and the aristocracy by the bourgeois class, the latter would ultimately be replaced by the proletariat and a classless society. In his thinking, Marx had evolved from the *dialectic idealism* proposed by G. W. Friedrich Hegel (1770-1831) to what he and his collaborator and friend Friedrich Engels (1820-1895) called *dialectic materialism*. This philosophy was supposed to describe the material relationship between labor and the objects it produces rather than an abstract Hegelian idea. Despite what Marx wanted it to be, this relationship was still an abstract idea.

Fig.44 Karl Marx (1818-1883) and Friedrich Engels (1820-1895) who as friends and collaborators together published the Communist Manifesto in 1848.

Early in 1848, the year of multiple revolutions in Europe, Marx and Engels shook the world by publishing their Communist Manifesto. This pamphlet openly laid out the aims of the new Communist League and encouraged workers everywhere to overthrow capitalist society, take over the reins of government, and enact Socialism. In some cases, they thought that capitalist society would collapse by itself from its internal contradictions (social injustice and unrest, cyclical depressions, unemployment, etc.), in others, as revolutionaries, they believed that it would require planning and the use of force. Having overthrown capitalism, they thought that the new regime would establish a *dictatorship of the proletariat* to take over the levers of political and economic power. Ultimately, a peaceful and egalitarian classless society would ensue and bring about universal bliss. As it turned out, this scenario didn't materialize anywhere during their lifetimes. Capitalism was able to adapt and bring about reforms that prevented its collapse. Marx realized this and went back to study the economics of capitalism and its inner workings. The result of these studies was his publication of the first volume of *Das Kapital* in 1867. Surprisingly, although Marx believed that capitalism should be overthrown, he also expressed admiration for the dynamism of the capitalist class towards urbanization, industrialization, productivity growth, and scientific and technological progress.

While unsuccessful in Western Europe, Marx and Engels looked eastward toward Russia. Their class-struggle model had not panned out there in the sense that the Decembrist Revolt of 1825, an uprising against the czarist government by the bourgeoisie, had failed. This gave the two of them the idea that maybe this historical stage could be skipped and that Russia could make a direct transition from a quasi-feudal regime to Communism. In this they turned out to be right, although it didn't happen until after their deaths: in Russia through the 1917 revolution under Lenin; in Eastern Europe by sheer force after the Russian occupation in 1945 under Stalin; in North Korea in 1948 under Kim Il-sung; later in China in 1949 under Mao Zedong; in Vietnam (1954-1975) under Ho Chi Minh; and in Cuba in 1959 under Fidel Castro. Marx and Engels' legacy was momentous but did not meet with success in the end.

In the West, capitalism survived the 19th and 20th centuries, but it did not do so without some opposition and corrections. The opposition came from multiple socialist movements, union movements and social-democratic governments that proposed a wide variety of economic reforms. Some thinkers believed that social justice would not be achievable unless the state took over all means of production, arguing that production is a public good that should be owned by all of society. Others were more moderate, believing in partial nationalization of industry while retaining much of private property and a multi-party democratic system. Meanwhile, the nagging problem of cyclical recessions and unemployment did not go away, and reared its very ugly head with the Great Depression of 1929. Towering over most economists, it was the young John Maynard Keynes (1883-1946) who started his active career in the U.K. around 1910 and eventually took the bull by the horns.

Fig.45 John Maynard Keynes (1883-1946) who revolutionized the field of economics in 1936 with his "General Theory of Employment, Interest and Money."

Between 1910 and 1929, Keynes began to publish controversial articles that enunciated two anti-establishment positions. The first was his condemnation of the excessive reparations imposed on Germany by the Treaty of Versailles. He reasoned that Germany was not in a position to pay for them and that they would bring about social unrest and political disaster, which they did. The second was his recommendation that the gold standard be dropped in the U.K. because it was only stifling the recovery of the British economy. His pro-stimulus recommendations during the 1920's, however, were mostly ignored until the Great Depression hit the world. In 1936 Keynes published his magnum opus, *The General*

Theory of Employment, Interest, and Money. Convinced that **the lack of demand** was the root cause of the depression (**and not the lack of supply**), he predicted that the market economy would not recover by itself without government intervention, as was assumed by so-called neo-classical economists. His proposed remedy was to lower interest rates and to spend funds on public works even if the government had to do it by deficit spending. Hitler had already figured that out by himself by rearming. Unfortunately, Keynes suffered a serious heart attack in 1937 which hampered his activities, and the Western democracies did not fully adopt his recommendations until WWII. When the governments were finally forced to borrow money and raise taxes to pay for the war effort, people were put back to work and the economic slump ended. After that, the *General Theory* gained enormous prestige in modern economic circles all over the world.

After WWII, Keynes represented the U.K. in the international monetary negotiations in Bretton Woods. His idea was to create strong institutions with rules that would manage international trade and prevent countries from incurring large trade surpluses and deficits. In this he was opposed by the more conservative American negotiators and he did not prevail. However, the World Bank, the International Monetary Fund (IMF), and the General Agreement on Tariffs and Trade were then created as a compromise which still greatly pleased Keynes. He died in 1946.

Starting in the 1960's, Keynesian economics fell out of favor for a while, in part because of the rise of economist Milton Friedman's libertarian ideas and the appearance of **stagflation** (see below). Friedman (1912-2006) published his *Monetary History of the United States* in 1963 and strongly opposed Keynes' idea that government should intervene in the market, other than by adjusting the monetary mass in circulation by printing more or less money. He had concluded, controversially, that the entire New Deal was "the wrong cure for the wrong disease," which he blamed on monetary contraction by the banks and poor policy by the Federal Reserve. Friedman also gained prestige by predicting "stagflation" in the U.S. and the U.K. in the 1970s. This phenomenon didn't seem to precisely fit Keynes' model because it allowed economic stagnation and unemployment to co-exist with inflation. Other economists explained stagflation by the oil crisis of 1973 which increased all production prices, and the failure of the U.S. government to raise taxes to pay for the Vietnam War. In the 1980s, Friedman was an advisor to President Reagan. His monetary policy combined with his abhorrence of "Big Government" fitted the Republican ideology like a glove and remained fashionable for a few more years. Eventually, Friedman's monetary theory failed to explain some new developments and lost some favor and influence. By the time the Great Recession hit the world in 2008 and banks began to fail, Keynes was making a strong come back.

In summary, four conclusions are apparent to me:

The first pertains to the fate of Communism and to some extent of Socialism. Had Marx and Engels been alive to witness the brutal dictatorships and gulags that followed the various Communist revolutions and wars, they might have regretted that their ideas unleashed this extreme violence and were used to justify it. Who knows? In any case, the blissful egalitarian society they expected never materialized. The top-down "command economy" planned by the state in the Soviet Union did not produce the desired prosperity, and Adam Smith's "self-interest," instead of leading to collective sharing of wealth, degenerated into bureaucratic corruption. In 1993, Communism collapsed in all the Soviet Republics and in Eastern Europe. The only country where a milder co-op system had been adopted was Yugoslavia, but that also collapsed when the country fell apart after President Josip Broz Tito's death. In China the Communist Party has so far kept control of political power but opened much of its economy to a market system. What exactly will happen to the economies of Vietnam, Cuba, and North Korea remains to be seen. Meanwhile, essentially all other countries in the world, including the U.S., have adopted some socialistic policies, whether explicitly or tacitly. These policies include various forms of safety nets for the poor, children, aid to public education, minimum wages and safety protections for workers, healthcare with differing degrees of public subsidies, financial security for seniors and handicapped, and so on. The devil is in the details: left-leaning politicians want more (cf. Bernie Sanders), right-leaning politicians want less, but the principle is no longer in question. This said, most of these countries have given up on the idea of taking over the means of production, one of the pillars of the Communist creed.

The second point is that there is no agreement yet among economists on how to avoid periodic booms and recessions in market economies. Booms are generally welcome and governments hesitate to stop them or even recognize them, although they have the tools to do so. Recessions are never welcome, and they still seem intractable. Ideological tension in governmental and political circles persists between two camps: those that believe that in a recession the remedy is massive deficit spending to increase demand, and those that think that the only solution is to adopt belt-tightening austerity until the system "cleanses itself from its sins." As a result of this tension, deficit spending, when implemented, tends to be too timid and less than effective. On the other hand, austerity is socially painful and very slow to bring about recovery, if it does at all.

The third conclusion is that capitalism and the market economy, which indeed are very efficient at allocating resources and promoting inventions and progress, are not good at preventing large income inequalities and poverty. Academics like Thomas Piketty in France and Robert Reich in the U.S. both show that wealth concentration can be inherent to the system but ultimately destructive of it.

Wealth concentration can be partially corrected through some combination of progressive taxation and a more generous social safety net, but right-leaning politicians generally oppose such measures. Meanwhile, most economists think that growth is indispensable to "lift the ship up" for everybody. However, we may reach a point in the not-too-distant future when growth will no longer be possible because of environmental limitations, and robots will replace part of the working class, thereby causing even more inequality and poverty. We are ill-prepared for these circumstances.

The latter two conclusions illustrate the two important conundrums that macroeconomics faces today. In addition, there is a fourth serious problem at the intersection of macroeconomics and microeconomics: the enormous power of domestic and multinational corporations, including banks and insurance companies. Corporations are the engines of the market economy. As Adam Smith stated, the "invisible hand" makes them provide us with an amazing wealth of goods and services in a decentralized way. There is no overall top-down management as long as free competition prevails. Corporations produce much of what we need, and also what we don't necessarily need. They treat their employees only as well as required to keep them satisfied and productive. But the success of corporations is based on the prospect of profits and growth because that is why people invest in them. Some corporations are more socially conscious than others but this behavior is just frosting on the cake, not their primary purpose. As corporations grow, they create powerful lobbies to push for favorable regulations and subsidies, or to defeat measures that might thwart their ambitions and their profits. They collude, become oligopolies and even full monopolies. They use devious means to deny that they might be harming people with tobacco, sweeteners, lead poisoning, pollution, and industrial hazards; they manufacture guns by the millions, can subject workers to poor working conditions, and so on. Together with wealthy citizens, they buy favors from politicians and fund their elections. They do everything they can to minimize their taxes and do not hesitate to move to countries with lower tax rates and cheaper labor.

Society has some defense mechanisms against the negative behaviors of corporations and large banks: smart and incorruptible legislators to enact effective regulations, strong labor unions; citizen's watchdog groups, and sometimes the press. It's often a battle of David against Goliath! This discussion leads directly to the next section.

Political Science

A review of the social sciences would be incomplete without mentioning political science. As a field of inquiry, political science is a relative latecomer to the humanities, having separated itself from political philosophy around the end of the 19th century. Its central purpose is to study systems of government, how they function, and where they derive their political power. It involves theories and models but generally cannot test them nor make predictions as can a physical science. Indeed, political systems and relationships change constantly and are affected by extraneous historical, legal, economic, and technological developments. Besides, political science is not a single discipline; rather it comprises several specialties from international relations, political economy, and the study of domestic public policies and administrations, political parties, and electoral systems.

There are currently about 196 independent countries in the world with various types of governments and electoral systems. One would think that even though political science does not easily lend itself to experimentation, comparing the successes and failures of these many forms of government might tell us something about their relative merits. Hence, if a country were fairly successful in running its affairs and satisfying the needs of its population, in theory it might be advantageous for another less successful country to adopt its governmental structure, constitution, and institutions. Unfortunately, in the real world such "adoption" may not work at all because of differences in their respective histories, the availability of effective leaders, the levels of education of their populations, their states of economic development, the respective levels of corruption, the religious backgrounds of their citizens, etc.

If political science cannot be a fully descriptive and predictive science, one might have hoped that it could at least serve as a form of prescriptive engineering, *i.e.*: "If you use such and such a form of government, then you might be able to expect the following results." But even such a modest expectation is unwarranted. Just take a look at the world today. With minor exceptions, most of the American countries from the U.S., Mexico, and all the way to Argentina have similar republican forms of governments with presidential regimes and separate legislative and judicial powers. And yet, all of them function differently and have their own difficulties and upheavals. Even the most stable and prosperous of them, the U.S., can elect a dismal populist leader and be paralyzed by gridlock and political polarization. The European Community, as a union with its assembly of parliamentary systems, is having difficulties in managing its common administrative and economic affairs and staying together. Witness the recent Brexit crisis, the rebellion in France, and the various tensions resulting from enormous flows of migrants. We don't know if and how Russia and China will evolve from their current authoritarian regimes. The former Soviet Republics, Pakistan, Afghanistan, and the African nations are struggling. India seems to be hanging on to democratic rule but it is not clear how it will emerge from poverty with its enormous overpopulation problem. The Middle East is in a continuous state of upheavals and religious conflicts. Bangladesh, Myanmar, Thailand, Malaysia, Vietnam, Indonesia, and the Philippines all have their own unsolved problems. North Korea is an absolute dictatorship. This leaves us with a few relatively stable countries like the U.S., Japan, South Korea, Australia, New Zealand, and Canada.

To understand how we got to this state of affairs, we must now look at history.

PART 4:

HISTORY AND LAW

As I mentioned in the introduction, our human condition unfortunately does not just hinge on the benefits of our knowledge, our education, and our command of science and technology. It also depends on our history, which is often ugly, bloody and brutal. History only happens once, it can never stop as long as humanity exists, and it is irreversible. As with science, its study relies on records, but unlike science, it cannot be tested.

History is the study of the past and it occasionally gives us glimpses of the future. It tries to describe and explain the lives of people, events, and the evolution of civilizations; the formation and dissolution of nations and empires; and the lives of their institutions and organizations. History can attempt to cover the entire world, *i.e.*, world history, or deal with specialized areas like national history, cultural history, military history, economic history, art history, gender history, history of science, etc. In all these domains, historians try to do two things: 1) gather information and validate it, and 2) discover causes and effects. Validation of collected information can take place only with the existence of records, documents, oral accounts, and the knowledge of traditions. Only a small fraction of all past events has actually been recorded. Accounts of some events are just anecdotal and may refer to events that never occurred. More recently, with the explosion of modern technology and 24-hour/day newscasts, events are being more frequently recorded, albeit on short-lived media. In the U.S. for example, police are being required to wear body cameras. And yet, still just a small number of events will be covered. Discovering causes and effects is an even more complicated task to which I will come back later. In any case, the perspective with which the past is viewed changes continuously with time. What seemed important yesterday may no longer look crucial tomorrow.

History in Early Greece and China

To see how historians got started and what they talked about, it is instructive once more to go back to Ancient Greece and two of its first historians: Herodotus and Thucydides, pioneers in the field. They set the stage for many historians to come.

Herodotus of Halicarnassus (484-ca. 425 BCE) is considered the "Father of History." He remains highly admired today, yet also controversial because a number of subsequent historians questioned his sources and the veracity of some of his stories. In his only work, *The Histories,* he recounts the origins of the fifty-year-long Greco-Persian Wars (499-449 BCE) and the famous battles of Marathon, Thermopylae, Salamis, and Plataea. Sadly, wars, bravery in battle, and conquests seem to take priority over the descriptions of more peaceful and constructive events.

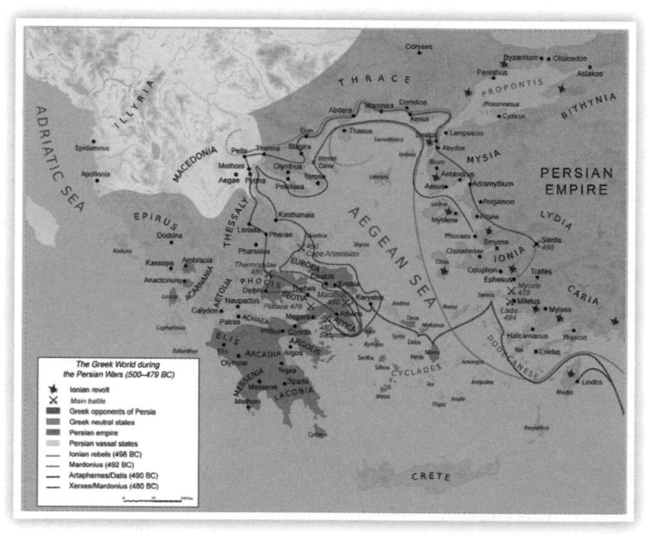

Fig.46 Map of the Greco-Persian Wars (499-449 BCE) as related by Herodotus (484-ca.425 BCE).

It appears that Herodotus gathered considerable information while travelling from Asia Minor to Samos, Athens, Egypt, and

Southern Italy. Like Homer, who drew on tradition and oral poetry, Herodotus told a number of tales and believed that the gods interfered to some extent in the lives of mortals. However, he did not believe in inevitable fate and systematically compiled materials and oral accounts to document his work and produce a coherent understanding of the sequence of events. It is said that he attended the Olympic Games in Greece and read his entire *Histories* to the assembled crowds in one sitting.

Thucydides (ca. 460 BCE-ca. 404 BCE) was brought up in Athens in a wealthy family that owned a gold mine in Thrace. In 434 BCE he became an Athenian military strategist and general. The war between Athens, a sea power, and Sparta, a land power, began in 431 BCE and ended in 404 BCE. In 424 BCE Thucydides was sent to Thrace to save the city of Amphipolis from the Spartans but he got there too late and failed. He was tried for his failure and exiled from Athens as punishment. For the next eleven years, he traveled all over Greece and eventually to Sicily, collecting information from all sides on the war which he published as his *History of the Peloponnesian War*[45].

His accounts of the events, in many of which he participated personally and took profuse notes, are related in meticulous detail. Unlike Herodotus, he did not believe in the intervention of the gods in the affairs of humans, and he was even-handed with respect to the various warring factions.

Thucydides starts out by describing the lack of unity among the various Greek city-states and examines the root causes of this bitter conflict that would last 27 years. He complains about the fact that, like barbarians and pirates, the Greeks everywhere were still carrying arms in peace time. He probably would not have supported a U.S.-style second amendment! The Athenians were finally the

45. See Richard Crawley's translation of Thucydides' "History of the Peloponnesian War," available online at http://classics.mit.edu//Thucydides/pelopwar.html

first to lay off their arms in their city-state. His factual prose, sometimes a little hard to understand, describes a geopolitical conflict in terms that today might be called *realpolitik*. *Realpolitik* is not a very precise term but it generally refers to the practices of leaders or writers who, both in peace and in war, believe that pragmatism should override ideology and morality. Machiavelli, Hobbes, and Henry Kissinger are often mentioned as more recent proponents of *realpolitik*. In this sense the Peloponnesian conflict was very similar to endless conflicts that have erupted worldwide in the subsequent 2500 years. It confronted two strong competing powers: Athens, democratic at least for some of the time; and Sparta, oligarchic and highly militarized. Both distrusted each other's motives, vied for influence with their neighbors, and created alliances: Athens with Corcyra (Corfu), and the Spartans with Corinth and the rest of the Peloponnesians. Tensions mounted as a result of several violent incidents, and the war finally became inevitable because of fears, mistakes, and interlocking alliances. Think World War I!

Fig.47 Spartan Hoplite Warriors.

To illustrate the points of view of the leaders on both sides, Thucydides quotes many of their negotiating positions, debates, and speeches. One of the most remarkable ones is Athenian leader Pericles' oration for his city's war dead. He opens his lengthy elegy by comforting the relatives of the deceased and praises their brave men for the contributions they made to the Athenian state and its superior democratic civilization. He then goes on to hail the dead at length for their courage and sacrifice and expresses the hope that the living will be inspired by their deeds. The speech sounds almost like it could be given today. Another wrenching passage in Thucydides' history is the description of the plague in Athens in 430 BCE. You almost feel like you were there!

Fig.48 Ancient Athenian Navy Ships.

Thucydides depicts the technical aspects of the Peloponnesian war campaigns very graphically. In 415 BCE, Athens made the disastrous decision to dispatch a large naval force to Sicily to conquer Syracuse, allied with Sparta, but its force was destroyed. For reasons unknown, Thucydides' account of the war stops abruptly in 411

BCE. Eventually, Sparta allied itself with Persia, fomented revolts around Attica, attacked Athens, and totally devastated the city. The Golden Age of Greece ended in much misery and destruction. Bad leadership! What else is new?

After reviewing how History was born in Ancient Greece, it is interesting to switch to China and the great historian Sima Qian (around 145-90 BCE). As author of the *Shiji* (Records of the Great Historian), Sima Qian covered more than 2000 years of Chinese history up to the Han dynasty under Emperor Wu during which he lived. At the age of twenty, Sima began to travel extensively around China and was able to create a detailed, first-hand account of life under the Han dynasty. After his travels, in 122 BCE, he was chosen to be a Palace Attendant in Xian and to travel with Emperor Wu to inspect various parts of the country. Following in the footsteps of his father, Sima Tan, he then became the court's astrologer around 109 BCE, by which time he was beginning to write history. Unfortunately, in 99 BCE, he got involved in the so-called Li Ling affair. Li Ling was a military officer who was sent by the Emperor to battle the Xiongnu people, considered barbarians. Li Ling was defeated and taken prisoner, for which the Emperor blamed him. As it turned out, Sima courageously defended Li Ling, and for this he was given a choice of one of three sentences: death, a fine he could not afford, or castration. Sima chose the latter sentence, became a eunuch, and was sent to jail. When he was freed three years later, he decided to return to history writing with one main purpose: to memorialize the lives of "good and worthy" people who may have suffered during their lives and that History might otherwise forget. In this he was probably inspired by his own fate, but his emphasis, at least, was not *realpolitik*!

Sima had a great admiration and respect for Confucius, and followed his advice to write only about what he knew, leaving out whatever could not be documented. His work extends over 130 chapters, covering not only many biographies but also astronomy, music, calendars, ceremonies, religion and economics. Frequently, he added his own judgment regarding the extent to which individuals lived up to traditional Chinese values of humility, self-discipline, and respect for the elders and the less fortunate. His style and approach had great influence in Korea, Japan and Vietnam.

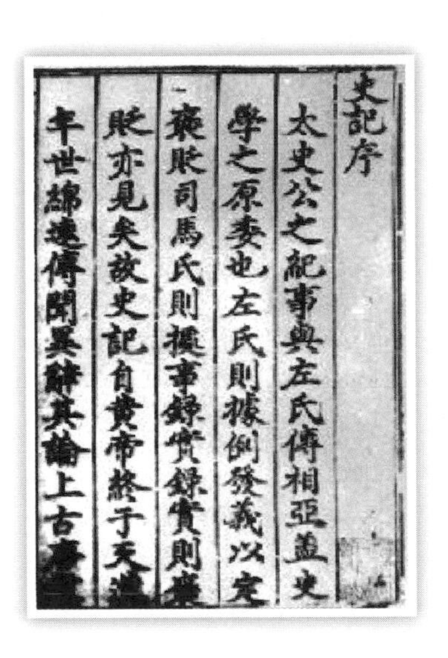

史記序
太史公之紀事與左氏傳相亞蓋史
學之原委也左氏則據例發義以定
襃貶司馬氏則撮事錄實錄實則襃
貶亦見矣故史記自黃帝終于天漢
年世綿邈傳聞異辭其論上古

Fig.49 The first page of the "Shiji' in which the great Chinese historian Sima Qian (~145-90 BCE) covered 2000 years of Chinese history.

History and Leo Tolstoy

Since the days of Herodotus, Thucydides and Sima Qian, there has been considerable discussion about what historians should study. It is a question with which Leo Tolstoy (1828-1910) struggled at length at the end of his novel *War and Peace*, and which Isaiah Berlin (1909-1997) summarized in a whimsical essay[46]. Tolstoy hoped that History could be made into an exact science but the more he thought about it, the more he concluded this was impossible. He attacked the question from several angles and ran into contradictions. The central one was that for History to become a tight causal Laplacian-style science, the actions of all the players would have to be driven by what he called *the law of necessity, i.e.*, determinism. But human consciousness implies that we have free will which, in his mind, conflicted with determinism. So he left this discussion open without final conclusion. He also never found an over-arching principle that would explain all of History, which is not surprising to me. I will come back to these questions of causality in History when I discuss the Cuban Missile Crisis below.

What Tolstoy did, however, was raise some important questions and be critical of past and contemporary historians. *E.g.*, isn't it a mistake to have historians concentrate only on the influence of leaders like Napoleon and Tsar Alexander I who impose their will and way of life on the masses? Aren't leaders in turn affected by their social and political environment and/or by the masses that transfer their power to them? What is this power and what forces does it depend on? Are these forces driven mainly by ideas? Would Napoleon have made war on all of Europe if he had appeared in the absence of the French Revolution and its ideas? But how could all the noble ideas of the French Revolution lead to twenty years of killing, first inside France and then all over Europe?

My personal observation, not entirely original, is that when historians write about broad trends and events, these relate inevitably to the formation and interaction of groups. Wherever groups of humans find themselves in any part of the world, they tend to

46. See Isaiah Berlin's "The Hedgehog and the Fox," Second edition, Princeton University Press, 2013.

congregate and eventually organize themselves into manageable units, bands, tribes, city states, nations, and empires. Their chiefs and rulers, driven by ambition, power, and the need for resources, have rarely hesitated to use conquest to aggrandize their territories and consolidate their hold over them. This has been true for the Greeks and Persians described above; the Pharaohs of Egypt; the emperors of the Roman Empire; the monarchs and emperors of Europe; the Tsars of Russia; the Mughals of India; the emperors of China and Japan; the emperors of Persia; the Caliphs and Ottoman emperors in the Middle East; the Aztecs, Mayans, and Incas in the Americas; and the Americans on the Frontier. Rulers of the past almost invariably managed their external policies by *realpolitik* and felt unconstrained by any international laws or ethical considerations. Many of their city-states, nations and empires rose and fell, some morphed into other forms, and some have survived up to our time. Their respective fates were determined by cultural, military, technological, and economic factors. Historians like Edward Gibbon (1737-1794), Oswald Spengler (1880-1936), and Arnold Toynbee (1889-1975) have written and studied their evolutions extensively. The quilted geographical map of the world today is the result of their multiple struggles and machinations.

The Invention of Democratic Representative Government

What is the most **stable, just, and effective** form of government? History has tried to answer this question for a long time.

The peripatetic Greeks in Athens first came up with a democratic system around 500 BCE. Their system was one of direct democracy, not representative democracy. Athens at the time had a total population of about 250,000 people. Participating citizens voted directly on legislation and executive bills. To vote they had to be adult male citizens and land owners. Women, slaves, and foreigners could not vote. As a result, only about 10% could participate. The government consisted of three parts: 1) an Ecclesia (Assembly), a sovereign body where at least 6000 of these people wrote laws, dictated foreign policy, and voted at any one time; 2) a Boule (Council) of 500 people representing the ten tribes of Athens that watched over the Ecclesia; and 3) separate Dikasteria (Courts) in which citizens could argue their cases before lottery-selected jurors. Athenian democracy flourished but did not last for more than two hundred years in its original form. Plato and Aristotle were highly critical of it because, like many others after them, they saw the system, particularly the Assembly's use of power, as the rule of the impulsive, uneducated poor that plundered the rich. Some Republicans in the U.S. Senate today might share these feelings! And yet, 2500 years later, this early Greek democracy, with all its weaknesses, slavery, misogyny, and wars still looks like an amazing political invention.

The other city-state in Antiquity that for almost five hundred years adopted a representative form of government was Rome. The Roman Republic (from *Res publica Romana*) began with the overthrow of the Roman Kingdom in 509 BCE and ended in 27 BCE with the advent of the Roman Empire when Augustus became emperor. During this period, Rome expanded its control from its city's surroundings to the entire Mediterranean basin, including France, Spain, North Africa, the Middle East, and Greece.

Roman government was headed by two consuls elected annually by the citizens, and advised by a Senate composed of appointed magistrates. Early on, the Senate was dominated by the patricians, Rome's land-holding aristocracy. Over time, the more numerous plebeians, or commoners, gradually gained power, became full members of the aristocracy, and, at some point, formed the separate Plebeian Council. Their accession to power was not smooth but they eventually came to learn to cooperate with the patricians and developed a strong tradition of public service, both in peace and in war. Interestingly, some of the laws enacted under the Republic and the subsequent imperial period made their way into the Justinian Code under Byzantine Emperor Justinian I between 529 and 534 AD. The Justinian Code, in turn, served as a model and template for the Napoleonic Civil Code issued in 1804 in France.

Most other nations and empires were ruled by autocrats who did not delegate their power. They may have surrounded themselves with courts and advisors but they ruled arbitrarily and for their own benefit. Despite this, they had to enact systems of laws to govern the behavior of their subjects or citizens, legal institutions to enforce them, and a bureaucracy to raise taxes. Codes of laws appeared almost simultaneously with early civilizations in Egypt (3000 BCE), Sumer/Babylon (2200-1760 BCE), in the Old Testament, in India, China, Japan, and Islam.

Fig.50 Babylon's King Hammurabi's code of laws.

Magna Carta and Common Law

It wasn't until the Middle Ages in 1215 that a seminal development took place in England. It was a peace treaty between King John and his barons, enshrined in what became known as the Magna Carta (Great Charter). Their conflict had arisen over mutual rights and taxes raised by the monarch who considered himself above the law. It went on for over eighty years, the reigns of three successive kings, generations of barons, and Popes Innocent III and Clement V. The final version of the Great Charter was reissued by King Edward I in 1297. Despite the fact that the document did not protect the rights of ordinary people, and many of its contents were subsequently repealed, it is still admired today by Anglo-Saxon legal experts and considered one of the greatest constitutional documents of all times. Despite the many conflicts and a temporary overthrow of the monarchy that took place in the

British Isles over the next few centuries, it lent them an amazing stability and resilience. It was the beginning of the codification of common law, and it eventually led to a parliamentary system of representative government under a monarchy. It is interesting that the U.K. after that has never enacted a formal constitution.

Today, one third of all nations, including the members of the British Commonwealth and the U.S., have inherited systems of common law from the British. These are based predominantly on jurisprudence and precedent, and some statutes. These countries have the longest traditions of representative government and the least amount of internal upheavals. Many of the other nations are ruled by systems descending from Napoleonic Law which is based predominantly on statutes, and sometimes on precedent. The rest are systems imposed by Communist regimes or Sharia Law (adopted in many Muslim countries). The latter two are not democratically chosen. They are enforced by authority or religion and are unlikely to satisfy their populations in the long run. Their futures are unstable.

None of the constitutional and legal systems described above are able to create perfect societies. Countries that have long-tested constitutions appear to benefit from greater stability but, inevitably, constitutions are products of their times and circumstances and can fall out of date. The U.S. Constitution of 1789, together with its original ten-amendment Bill of Rights, was an admirable document but it was not able to prevent the Civil War. It has been amended seventeen times since its inception. The Thirteenth Amendment banning slavery came too late. The Second Amendment giving the people's militias the right to bear arms was drafted to protect the new country from the British. It has become obsolete but cannot be repealed because it has now morphed into a right to defend oneself against one's fellow citizens. As a result, today there are three-hundred million guns and assault weapons floating around the U.S. and raising havoc. Is that good? The U.S. Supreme Court's so-called originalists and reformers are split on this and many other issues.

Having said this, France, since its 1789 Revolution, has had about sixteen constitutions, three monarchies, two empires, and five different republics. Its history has been very unstable. But which was the cause of the instability: the lack of an appropriate form of government or the inability of the French people to rule themselves? Has a new page of French history been turned with the election of President Emmanuel Macron in 2017? Not obvious!

International Relations and the Balance of Power

Unfortunately, the state of international law is even more unsettled and ineffectual than that of domestic law. When a crime is committed within a country, the police are called to get matters under control and the justice system is brought in to rule on the matter. When an act of international violence is committed, there is generally no police force and no international law capable of stopping it. Such conflicts can fester for years.

The history of international relations is a checkered one: it shows some occasional progress[47] but it is not very encouraging. In Europe, one of the turning points was the Peace of Westphalia in 1648. Treaties signed in Osnabruck and Muster put an end to the Thirty Years War (1618-1648) in the Holy Roman Empire and the Eighty Years War (1568-1648) between Spain and the Dutch Republic. What was important about this event is that it estab-

47. A surprising international agreement that made its appearance around 1300 was the Lex Mercatoria or Merchant Law. This law, which was not imposed by governments, was a type of common law derived by merchants all over Europe to regulate their commercial interactions, property rights, contractual obligations and best practices. It was enforced by an agreed-upon system of merchant courts, and an amazing example of self-enforced international law. International trade flourished and states managed to raise large amounts of taxes from it. It was a precursor of the World Trade Organization (WTO).

lished the concept of sovereign states. To maintain and respect the sovereignty of each, these states were forbidden from interfering in each other's domestic affairs. This concept thereafter spread all over the world. Aggression between states was supposed to be prevented by a *balance of power* achieved through networks of alliances. This system still prevails today but in fact it has only intermittently succeeded. World Wars I and II are perfect examples of its failure.

Conflicts today can take three major forms:

The first is between two or more feuding factions and nations: *e.g.*, between South Korea, North Korea, Communist China, and a U.N.-led Coalition; Iran and Iraq; Israelis and the Palestinians; India and Pakistan; Argentina and the U.K.; Iraq and Kuwait (ended by the Desert Storm coalition); or Armenia and Azerbaijan. The United Nations and its Security Council were specifically created to prevent and stop such conflicts. Unfortunately, the Security Council only works effectively when the five permanent members (China, France, Russia, the U.K., and the U.S.A.) reach unanimity on a case and agree to intervene with diplomacy and possibly by force with peace-making and peace-keeping troops. By a fluke the Security Council's authority worked in Korea in 1950 because the Soviet Union was absent, boycotting the U.N. to protest the non-recognition of Communist China. But most of the time this doesn't happen and the Security Council is then paralyzed by the veto of one or more of its permanent members. I will come back to this problem in Part 5.

The second form of conflict is civil wars: *e.g.*, North and South Vietnam, Ruanda, the Congo, the former Yugoslavia, Afghanistan, Colombia, Libya, South Sudan, Syria, and Ukraine. In these cases, the U.N. Charter still works on the Westphalian principle. It states that the right of national sovereignty trumps any other contingencies (even genocide) and forbids the U.N. from intervening with force inside a country. This, by the way, does not prevent individual actors or coalitions from stepping in. For example, in the case of the Syrian civil war, when the Alawite government began to attack the Sunni population, the U.N. did not intervene. When the government started to use chemical weapons, an agreement was reached between the U.S. and Russia to have them removed. Apparently it did not entirely succeed. Meanwhile, the opposing parties called in military proxies from outside, and the war escalated to an unprecedented level of brutality. Nobody wanted to lose, but in the process the number of Syrians being killed and escaping the country kept increasing dramatically. During his tenure, U.S. Secretary of State John Kerry travelled all over the world to secure a truce but the U.N. was powerless. As this book is being completed, military might seems to have overpowered right.

There is a third type of international conflict that the U.N. seems to be unable to stop. It is when a superpower decides to take matters into its own hands. This type includes China's intervention in Tibet; the temporary Soviet takeover of Afghanistan in 1979; the U.S. invasion of Iraq in 2003, which destabilized the entire area; and the recent Russian takeover of Crimea from the Ukraine. The U.S. intervention in Afghanistan in 2001 in retaliation of the Al Qaeda attack on the U.S. on 9/11 is understandable. However, even in this case, if the U.N. could have functioned more effectively, it might have mobilized the entire world community rather than leaving the matter in the hands of a coalition of NATO members.

The Balance of Terror: the Cuban Missile Crisis and Vasili Arkhipov

As serious and unpredictable as the international situation is today, there is one factor always looming in the background that makes it much worse: the risk of nuclear war either by intent, miscalculation, or through a terrorist plot. We have moved from a balance of power regime between sovereign states to a balance of terror regime between nuclear states. For this reason, and partially motivated by Tolstoy's futile attempt to find predictable causes of world events, I want to review and summarize a 20th century historical crisis that took place in my lifetime in which all humanity could have been destroyed: the Cuban Missile Crisis. It must not be forgotten because it was terrifying, and a crisis of similar gravity could be triggered again in the future in some other part of the world, such as in Korea or between India and Pakistan.

The Cuban Missile Crisis of 1962 had many independent causes and antecedents. One antecedent was the Cold War and the rivalry between the Soviet Union and the U.S., an antagonism with endless causes. A second was the invention and existence of nuclear weapons and delivery systems, an immediate result of WWII but with endless other causes. A third one was poverty and exploitation in Latin America, specifically in Cuba. The latter led to the Cuban Revolution of 1959 and the overthrow of Fulgencio Batista by Fidel Castro, with the help of Che Guevara and a revolutionary uprising. A fourth cause was the lack of understanding of the roots and meaning of this revolution by the Eisenhower administration, which saw it only as a Communist threat, and hence hatched their CIA plan for the Bay of Pigs invasion. A fifth one was the victory of John F. Kennedy over Richard Nixon in the 1960 presidential election. Note that this victory itself also depended on many factors and idiosyncratic details, some ideological, some financial, some other: JFK's campaign support from his father; JFK's intelligence, youth, charm, and humor; Nixon's five-o'clock shadow and uneasy affect during a televised debate; the promise of a "New Frontier;" JFK's incorrect but effective argument about a "missile gap" (that really didn't exist); and so on.

On April 17, 1961, shortly after JFK became president, he authorized the CIA-sponsored paramilitary Cuban Brigade 2506 attack on Cuba at the Bay of Pigs. Personally, having lived in Latin America for some time and knowing about the socio-economic conflicts there, my immediate reaction was: "This is a huge mistake and it's going to fail." And so it did. JFK was poorly advised, and as smart as he was, he did not cancel the CIA plan plotted under Eisenhower's administration. By then, Castro was irreversibly falling into the Soviet Communist sphere of influence. Knowing that another U.S.-plotted attack was coming sooner or later, Castro sought Soviet protection. In a secret meeting around May 1962, Krushchev promised Castro that the Soviets would construct a number of nuclear missile launch sites in Cuba pointing at the U.S., 90 miles away. The camouflaged construction took about four months, after which the first missiles were delivered. It took some U.S. reconnaissance effort before the sites could be photographed by an American U-2 spy plane on October 15th, 1962. The crisis then immediately escalated.

Kennedy met with his advisors and military strategists to carefully examine all his options, from doing nothing to bombing and invading Cuba. On October 22nd he broke the news to the American people and to the entire world, and declared a blockade on the Soviet fleet to prevent delivery of further missiles. Very intricate negotiations involving several exchanges between Kennedy and Krushchev and various diplomats followed, including consultations with allies and Latin American countries. U.S. nuclear missiles and bombers were put on fifteen-minute alert.

There were two very serious military incidents during the crisis. The first involved the incursion into Cuban airspace of another U.S. U-2 plane which the Russians shot down, killing the pilot. The Americans had in principle decided that this would constitute a *casus belli,* but at the last minute they chose to ignore it, not being sure of who in the Soviet hierarchy had given the orders to shoot it down. The second incident happened when the U.S. Navy damaged a Russian B-59 diesel-powered submarine at the blockade line with depth charges, unaware that it was armed with a nuclear-tipped torpedo. There were three senior Russian officers in charge on board: Captain Valentin Grigorievitch Savitsky, the political officer Ivan Semonovich Maslennikov, and the second-in-command Vasili Alexandrovich Arkhipov, who by luck had also previously been flotilla commander. They had lost contact with Moscow and didn't know if war with the U.S. had broken out but they had orders that if the submarine was "hulled" (pierced), they should launch the nuclear torpedo under the condition that all three officers would agree. Savitsky and Maslennikov agreed but fortunately, Arkhipov objected, so the launch did not take place. It may well be that Arkhipov, an unsung hero, averted World War III.

Fig.51 Vasili A. Arkhipov (1926-1998), the man who may have saved the world from nuclear war.

On the same day of this incident, on October 27th, an agreement was finally reached between Kennedy and Krushchev. The Soviets would dismantle and remove all their intermediate-range nuclear missiles from Cuba (and eventually also remove their tactical nuclear missiles, not even known to be there by the Americans). The U.S. made two important concessions as well: 1) commit to never support an invasion of Cuba again, and 2) remove its Jupiter missiles pointing at Russia from Italy and Turkey. Eventually, a Moscow-Washington communication hotline was established to reduce tensions between the two countries in the future. A year later, JFK was assassinated, and in 1964 Krushchev was deposed.

This summarized account of probably the most serious crisis of the Cold War shows that historical reality in all its complexity can sometimes be reconstructed with enough time and witnesses, but that no scientific model or computer program could have predicted its onset and its outcome. Causal factors operate at too many different levels to enable anyone to create a deterministic network. What can be said *a posteriori* is that in this "balance of terror" situation, both Kennedy and Krushchev weighed the consequences of their options, and in the end made somewhat rational decisions to avoid a catastrophe. That, to answer Tolstoy's central question, is as close as leaders can come to using their *free will*. But despite this, the catastrophe almost got away from them. If history ever teaches us a lesson, this should be it.

The Future of Nuclear Weapons

The lesson is actually very simple: we should get rid of all nuclear weapons, worldwide. But how do we do this? Nuclear weapons create horrendous fears but they are also built out of fear. From the beginning, the U.S. used the Manhattan Project during World War II to build the atomic bomb out of fear that Hitler would build and use one first. Hitler in the end did not succeed. Eventually, in August 1945, the U.S. dropped two atomic bombs on Japan, the first on Hiroshima and the second on Nagasaki. The war ended almost instantly, but the Russians, not wanting to be left in an inferior military position, started their own nuclear bomb program. The British, French, and Chinese followed suit. The Nuclear Arms Race took off with a vengeance. Hydrogen fusion bombs first supplemented and eventually outnumbered nuclear fission bombs. Delivery systems escalated from bombers to ICBMs and submarines, and battlefield tactical nuclear bombs were developed and deployed. At the height of the Cold War, there were more than 68,000 "active" nuclear weapons on earth, enough to obliterate life on the planet several times over. Adversaries realized (not always clearly) that these weapons, some of them a thousand times more powerful than the two dropped on Japan, could not be used to fight wars, but only as deterrents in a balance of terror regime called MAD, *i.e.*, Mutually Assured Destruction.

Following the Cuban Missile Crisis in 1962, four bilateral treaties were negotiated by the U.S. and the Soviet Union: the Partial Test Ban Treaty (PTBT) in 1963 banning nuclear tests in the atmosphere, oceans, and space (but not underground); the Strategic Arms Limitation Talks (SALT I) in 1969; the Anti-Ballistic Missile (ABM) Treaty in 1972; and SALT II in 1979.

The INF Treaty (Intermediate-Range Nuclear Forces), which was signed by President Reagan and Soviet leader Gorbachev in 1987, eliminated all nuclear and conventional missiles with ranges of 500–1,000 kilometers (short-range) and 1,000–5,500 km (intermediate-range). In 2014, the U.S. formally accused Russia of a treaty breach for developing a prohibited ground-based cruise missile. The dispute has not yet been resolved and threatens the future of the treaty. President Trump is threatening to withdraw the U.S. from the treaty.

By the late-1990s, the U.S. and Russia had negotiated and concluded the 1991 Strategic Arms Reduction Treaty (START I) and the 1993 START II. The 2010 New START Treaty took each nation's strategic deployed arsenal down to 1,550 warheads by

2018. That total is still 1/3 larger than the Pentagon has determined as necessary for U.S. nuclear deterrence needs. New START will expire in 2021, unless extended by mutual agreement.

In addition to these bilateral treaties, two other multilateral treaties were negotiated:

The 1968 NPT (Nuclear Non-Proliferation Treaty) bars non-nuclear weapon states from acquiring them and urges nuclear-armed states to end the arms race and achieve nuclear disarmament. This treaty to date has been signed by 191 nations. Those who have not signed include India, Pakistan, and Israel. North Korea is in breach of its NPT commitments.

The 1996 CTBT (Comprehensive Test Ban Treaty) prohibits all nuclear test explosions, thereby creating a barrier against new nuclear warhead development, to date ratified by 166 nations and signed by 183 states. The U.S. and seven other states must still ratify the CTBT for it to enter into force.

Despite all these treaties and many attempts by peace organizations and world leaders, including Henry Kissinger, William Perry, George Schultz, Sam Nunn, and Sidney Drell[48], there are still about 15,000 nuclear warheads in the world, many on active alert:

48. See Philip Taubman, "The Partnership: Five Cold Warriors and their Quest to Ban the Bomb," Harper 2006.

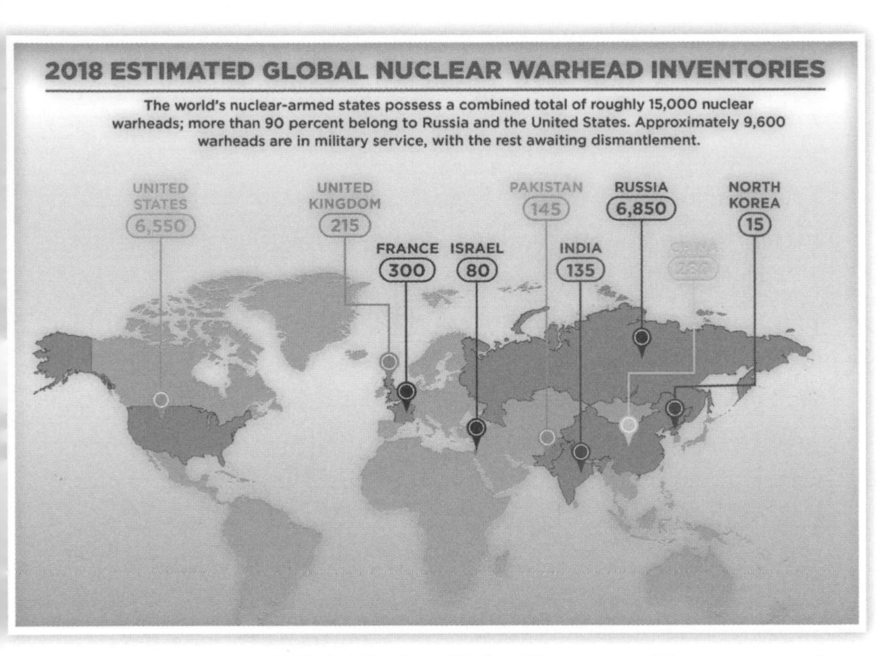

Fig.52 2018 Estimated Global Nuclear Warhead Inventories. Ninety percent of these roughly 15,000 warheads belong to the U.S. and Russia. Approximately 9,600 warheads are in military service, with the rest awaiting dismantlement.

The risk of nuclear weapons getting used some day continues to threaten all of humanity

Conflicts like the Cuban Missile Crisis can re-occur and rational or irrational leaders can still start a nuclear war by miscalculation. Serious nuclear accidents and mistaken launch orders have occurred in the past and will probably happen again.

Of the nine nuclear-weapon states, only two—China and India—have adopted a "no first use" policy.

A significant number of the nuclear weapons deployed by the U.S. and Russia are still on a "launch-under-attack" posture, meaning that the leaders of either country could order the launch of hundreds of nuclear warheads on ICBMs within a minimum of four minutes and on SLBMs within fifteen minutes. "De-alerting," *i.e.*, mutually agreeing to lengthen the above "short fuses" has been advocated for years but the U.S. and Russia have never agreed to this for fear that they would weaken their respective deterrence postures.

The use of just a portion of the U.S. and Russian arsenals, or a nuclear exchange between India and Pakistan would produce catastrophic immediate consequence for hundreds of millions of people, with global climactic and fallout effects potentially affecting billions.

If terrorists get hold of a sufficient quantity of unsecured, highly enriched uranium or plutonium, they could build crude nuclear bombs and create havoc.

On July 7th, 2017, after several months of deliberations, 122 countries at a special U.N. conference formally adopted a **treaty** that categorically prohibits the use of nuclear weapons and ultimately strives for their total elimination. Similar treaties had already been adopted in the past for chemical and biological weapons, land mines, and cluster bombs. Forty countries have refused to support the treaty for now. They include the nine nuclear weapons states, most other European countries, Canada, Japan, and Australia, all considering themselves protected by the U.S. nuclear umbrella. Much hard work lies ahead to eventually bring them on board.

PART 5:

THE FUTURE OF THE HUMAN CONDITION?

Making Predictions

Today is January 9th, 2018.

Six months ago, I thought I had written the final version of this section of my essay. But then a number of events transpired and I discovered Yuval Harari's two recent books (Refs. 1 on "Sapiens" and 2 on "Homo Deus"). As a result, I feel compelled to bring up some of his predictions and modify some of my conclusions.

By total coincidence, as I begin to revise this section, two young deer have appeared in front of my window facing my side-yard. This is a very rare occurrence and I have never been visited by two deer at the same time. I don't know if they are male or female but for convenience, I am calling them Tony and Mary. Tony is partially hiding behind some acacia branches and is eating grass. Mary is in plain view and she has discovered me, sitting in my study in front of my computer. Almost completely immobile, she looks at me intensely for almost ten minutes. Only her large ears seem to be wiggling back and forth, perhaps to catch some extraneous sounds. I move my head a little to see if she is going to react but she continues to just stare at me, as I do at her. I don't know what's going on in her head. Mary "is certainly not" an algorithm (as Harari conjectures) but she may be **partially** run by one. What I can say, in comparing our behaviors, is that we both seem conscious (aware) of each other, even if our states of consciousness are different. She stands still and observes my most minute movements. Unlike me, she doesn't have a computer keyboard to write about our interaction. Eventually Mary seems to lose interest and she slowly walks away. As she does, however, she turns her head once more to look at me (as if she were still "wondering" about me?) before disappearing behind the trees. Eventually, Tony follows her. Does she report to him that while he was eating, she met a *Homo Sapiens*? Can I figure them out? Can I predict their future or the future of the "deer species condition?"

After this unexpected interlude, I can come back to our human condition. As I mentioned in the introduction to this book, we seem to be stuck in a serious dichotomy. On the one hand, our species has triumphed in accessing, studying and using the reality of our world to our benefit. We have developed amazing sciences and technologies that enable us to live more secure, informed, and better lives than any of our ancestors. On the other hand, we so-called *Homo Sapiens* don't seem to be wise enough to organize our societies in a rational, ethical, and peaceful manner. At the individual level we fight and mistreat others, and collectively we continue to displace and kill millions of people in wars and revolutions. As a species we are gradually destroying our environment and millions of other species with which we share our planet. Is humanity doomed forever to live with this dichotomy? Can we predict the future? Can we influence it?

If you haven't been entirely convinced by my earlier argument that history is chaotic and unpredictable, let me just do a small experiment by turning the clock back fifty years to January 1968.

On a personal level, at that time I was a bachelor. Could I have predicted that in 1969 I would get married? No.

Could I have predicted that by 1974, I would have two children? No.

At the time, I had been working at the Stanford Linear Accelerator Center (SLAC) for ten years. Could anybody at SLAC have predicted that by 1969, the existence of sub-nuclear particles called quarks would be firmly established there and that three of my colleagues would get the Nobel Prize for the discovery? No.

Could I have predicted at the time that I would continue to work at SLAC for another forty years? No.

Could we have predicted that in 1969 humans would land on the moon and make it back to earth safely? No.

Could anybody have predicted that President Johnson would suddenly decide not to run for the Presidency, that Robert Kennedy would enter the race, and that he would be assassinated in June? No.

Could we have predicted that Richard Nixon would then be elected President in November 1968? Maybe, but that during his second term he would be forced to resign in 1974 after the Watergate scandal? No.

In France, could anybody have predicted that President Charles de Gaulle would resign the next year? No.

Could anybody have predicted the world oil/energy crisis of 1973? No.

Could anybody have predicted that the Vietnam War would not end until 1975? No.

A few scientists around 1968 were warning that we might be running into some environmental problems, but had anybody predicted that global warming caused by green-house gases would become a planetary threat fifty years later? Not quite.

Did any economist predict the astounding rise of China's economic power, the effects of globalization, and the dangerous increase in economic inequality worldwide? No.

And so on.

In other words, minor and major events cannot be predicted, even one or many years into the future.

How about predicting trends, if not specific events? How about considering the following approaches:

We may be able to study and discern certain **undesirable** trends on the horizon and try to prevent them from materializing ahead of time. This is the approach Harari has taken in his "Homo Deus" (ref. 2).

We may wish certain **desirable** trends to materialize but realize that they won't unless we act ahead of time to give them a chance.

This second approach is the one I am taking on as the basis of some of my observations in this book.

Sometimes these two approaches get mixed together.

To summarize what I said about studying reality, I stated that we live in an *objective real world,* independent of us and common to all of us. As Kant stated, we cannot exactly grasp it except that we constantly interact with it through our senses. Evolution has developed our senses which give us a *representative reality* of this world via our perceptions. Our brains which have a structure of their own, integrate all these perceptions, store many of them in our memories, and allow us to produce theories, histories, laws, models, maps and dreams that are part of what I have called our *mental subjective reality.* Using the scientific method, we record ever more precise observations and measurements, and develop ever more powerful models capable of predicting, and if possible explaining, the behavior of the objective world which includes us and our bodies. The product of this endeavor is what I have called *scientific reality.*

The Future of Human Evolution

To look into the future, we may begin by asking: what is the chance that human nature will change fundamentally? By "fundamentally" I mean beyond the changes we induce in ourselves with mood-modifying pills, opioids, and other drugs. Might our senses and brains continue to evolve? A catastrophe like the asteroid impact that may have caused the extinction of the dinosaurs about 65 million years ago, a nuclear war, or a worldwide incurable pandemic could change everything. Barring such events which could have large-scale epigenetic effects, and in the absence of human intervention, random genetic variations and mutations will continue to take place for humans. But will natural selection continue to occur for us? For the time being, birth control, universal access to medicine, and somewhat egalitarian policies will probably enable large numbers of people to survive and reproduce fairly uniformly. People with genetic advantages may be able to live longer but that doesn't mean that they will have more children to transmit whatever traits they carry. As a consequence, **natural** selection for humans is unlikely to continue on a significant scale.

On the other hand, what about "**artificial**" selection, *i.e.*, selection driven by us? The answer to this question is the one that preoccupies Harari the most. It is multifaceted and complex. To come to grips with it, let us start with non-human organisms. As early as 12,000 years ago, long before we discovered DNA and the workings of genetics, farmers already began experimentally to improve their crops and their domesticated animals through selective breeding. The process was relatively slow and safe. But all of this is rapidly changing now: the exploding science of genetics is expanding our options dramatically with the introduction of new knowledge and technologies such as GMOs and CRISPR.

GMOs (Genetically Modified Organisms) are organisms whose genome is artificially modified through deletion or addition of one or more genes of specific characteristics. If the addition consists of a gene from an unrelated organism, it is called "transgenic." The purpose of GMOs is to produce modified foods, medications, and other products, and to do research. It began with the fairly simple modification of bacteria. Some of these can now be used to produce insulin, growth hormones, and biofuels. Applied to agriculture, the technology tries to introduce a new trait into a plant, *e.g.*, to make it resistant to spoilage (tomatoes) or a disease (tobacco, potatoes), or to an insecticide or an herbicide (like Monsanto's Roundup used to exterminate collateral weeds). Farmers, particularly in the Americas and in developing countries, have massively adopted GMOs to increase the yields of crops such as soybeans, corn, and cotton. Applied to animals, the technology is used to study human diseases, improve animal health and accelerate growth for our selfish benefit (*e.g.*, for pigs, tilapia, and salmon), produce an enzyme to speed up the making of cheese, and modify dairy cows so that they produce milk identical to human breast milk.

GM foods are an excellent topic to bring up at a dinner table if the conversation is dying: it will instantly revive it by getting everybody excited. The reason is that in the absence of full information about the safety or risks involving GMOs, people feel entitled to strong opinions. The fundamental problem is that GMOs differ greatly from each other, and their effects need to be tested, case by case. Based on such tests, most of them indeed appear to be harmless, at least in the short run. But what about twenty years later, after the organisms containing these modified molecules have been dispersed into the environment and have invaded the entire food chain? How can we be sure? We obviously cannot. Continued research and vigilance are needed.

The legal status of GMOs varies greatly in the world, with some countries banning them altogether. In other countries, GM foods must at least be labeled, but this is not the case in the U.S. at this time. Requiring GM foods to be labeled gives the public the option **not** to use them but they have become so ubiquitous to be almost unavoidable. In this respect we certainly need to be skeptical of Monsanto's enormous market domination in the field. On the other hand, what should be done about the production and distribution of **golden rice?** This rice, modified to contain beta-carotene (a precursor of vitamin A) could save millions of children in Africa or Bangladesh suffering from Vitamin A Deficiency (VAD) and risking blindness or starvation. Greenpeace adamantly opposes golden rice but Pope Francis, one hundred Nobel laureates, and the Gates Foundation support it. How does society decide? We don't seem to have established ethical and legal guidelines in place.

Let us now move on to humans. As pointed out by University of Manchester's Prof. Matthew Cobb[49], we are currently witnessing two technical genetic revolutions. The first, fairly well understood, is that our individual genome can now be easily sequenced and tested for genetic diseases such as Down syndrome, Tay-Sachs disease prevalent with Ashkenazi Jews, and sickle cell anemia predominantly affecting African-Americans. Having this information is a positive development but it means that two members of a couple, both carrying this gene, can now practice what Cobb calls "soft eugenics" by deciding not to marry or at least not having children: "With the best of intentions and, for the moment the best of outcomes, we have drifted across a line in the sand." Furthermore, a new inexpensive blood test for Down syndrome on a pregnant mother means that, if positive, it will put pressure on her to consider an abortion, another difficult moral dilemma. Finally, in an era when health care plans and insurance companies are obsessed with pre-existing conditions, what will they do when the results of these genetic tests become public knowledge?

The second revolution has to do with the advent of genetic sequences to edit human, animal, and plant DNA. Of these, the so-called CRISPR sequence is the most recent, inexpensive, and powerful. It was accidentally discovered in 2006 in bacteria. These bacteria were found to use CRISPR ("clustered regularly interspaced short palindromic repeats") to store chunks of DNA of attacking viruses between these "repeats." The bacteria thereby learn how to recognize future viral attacks and develop immunity to them, like a vaccine. In 2013, scientists got the idea that they could associate CRISPR RNA sequences with a particular enzyme

49. See Matthew Cobb's article "The Brave New World of Gene Editing" in the New York Review of Books issue of June 13th, 2017.

called Cas9 to act like a molecular pair of scissors. The scissors can be guided to any desired DNA section of a genome by designing a CRISPR RNA sequence to match that of the DNA to be targeted. After cutting the DNA, two choices are available: one can either let natural mechanisms repair it if that is possible, or one can insert a new piece of DNA to modify it for a given purpose. In less than four years, the CRISPR technology has exploded to a point where it is widely used to produce genetic modifications in plants and animals, particularly mice and pigs. CRISPR may also be used in a much more aggressive and fast acting intervention called "gene drive." By this technique, it could for example modify the genes of mosquitoes so that, through rapid reproduction, they would become collectively inhospitable to the malaria parasite they normally transmit to humans, or be made sterile. This seems like an interesting idea, but what if the parasite mutates and becomes even more uncontrollable, or if the disappearance of all mosquitoes results in an unexpected harmful ecological surprise?

For humans, the big hope is to find ways to safely repair and edit multiple somatic cells responsible for inherited diseases, HIV, and certain types of cancer. Tens of millions of dollars are being invested in studies to achieve these purposes. Beyond this and far more controversially, CRISPR technology can be used to edit defective genes in single-cell embryos before they multiply. As this book is being written, this so-called **germline** intervention has already been tried on non-viable embryos in the U.K. and the U.S., and at least once on viable twin embryos in China to make them immune to AIDS. If used on viable embryos in the future, it carries the risk that whatever favorable editing it might produce, it could simultaneously cause unintended and dangerous DNA changes that would afflict the born individual and then be passed on uncontrollably to future generations. Not surprisingly, the CRISPR/Cas9 technology is raising profound medical, ethical and socio-economic issues.

There is a huge tension between the immense benefits that might be derived from it and the commensurate damages and abuses that might ensue. It requires that restraint be exercised by the scientific community, that society be educated on this issue, and that governments develop enforceable international regulations to prevent irreversible biological disasters. A comprehensive discussion of all these problems can be found in a recent book by Jennifer Doudna and Samuel Sternberg[50].

If the consequences of unleashing the CRISPR technology on humans could be momentous, they are nothing compared to the "psycho-technologies" that Harari (Refs. 1 and 2) thinks are already becoming available or at least are not far on the horizon, driven by the profit motive of the market place:

a) the potential that biologists will find ways of genetically engineering **enhanced** brains for elites that can afford them, leaving everybody else behind, perhaps without jobs,

b) the potential that biologists and computer specialists will be able to create brain-computer interfaces (BCI) between people's brains and computers that can pass thoughts, algorithms, and data back and forth between them (see Ray Kurzweil[51]),

c) by extension, the so-called cyborgs (cybernetic organisms) that result from combining living humans with electromechanical devices, either to restore lost functions (as a result of accidents) or to produce enhanced functions (extra intelligence or physical strength),

d) the possibility of extending the lives of humans indefinitely, making them immortal (but not necessarily immune to a car or plane accident!),

---

50. See Jennifer A. Doudna and Samuel H. Sternberg, "A Crack in Creation, Gene Editing and the Unthinkable Power to Control Evolution", Houghton Mifflin Harcourt, 2017.

51. See for example Ray Kurzweil, "How to Create a Mind: the Secret of Human Thought Revealed," Viking, 2012.

e) or beyond this, the potential of creating entirely synthetic inorganic sentient beings that will be able to replace *homo sapiens* altogether,

f) all of this combined with the computer studies carried out by Google, Facebook, Amazon et al. to track our behavior, our thoughts and our desires, and eventually connect us to a mega database and run our lives like puppets via a universal Internet!

If one or any combination of these Orwellian scenarios were to materialize on a grand scale, it would be a disaster and it would be the end of human life as we know it. I can't estimate what probabilities to assign to these scenarios, but Harari is doing us a huge favor by giving us a warning: if we don't watch out, this future might become our default condition. We must widely publicize all these threatening possibilities in the media and make sure to create worldwide institutions capable of blocking them.

To summarize, I believe it would be wonderful if we could rid humanity of horrible diseases and suffering. It might also be beneficial if we could improve our psychotherapy techniques. However, we should be extremely skeptical about tinkering irreversibly with our *representative reality*. It seems to me that we will be better off if we leave the ways we perceive the color and fragrance of a rose as they are, if Mozart continues to sound to us like Mozart, if chocolate never ceases to taste like chocolate, and if sex continues to feel as it does.

The Development of the Physical Sciences and Technology

Having said this, I don't mean to imply that we humans should stop being curious and inventive and worry about our material survival. I believe we should and will continue to push the boundaries of what I have called *scientific reality*. This is one of the beauties of the human enterprise and spirit, and we should not abdicate it. By pursuing it, our understanding of the universe will increase; we may someday figure out if the Big Bang was unique or if there are other universes out there; we may discover the nature of dark matter and dark energy, and whether one of the so-called string theories described by Stephen Hawking (ibid ref. 3) has any experimental validity in physics; in biology we may eventually discover and understand the origin of life; through genetics and medicine as mentioned earlier, we will be able to cure diseases so far incurable; through energy research, physics, chemistry and the geosciences we may find ways to slow down global warming and its consequences, and also to predict earthquakes. Furthermore, even though the fundamental constituents of matter will almost surely remain unchanged, the *objective real world* of which it is composed will never sit still: the universe will continue to expand in the foreseeable future, new galaxies will form and merge, stars will be born and die, new volcanoes will erupt, tectonic plates will drift, species will come and go, and humans will continue to manipulate and modify their environment. Hence, we will always have something new to observe and study and we will, no doubt, continue to greatly improve our *physical map* of the world. This is true even though Kant and Korzybski may be have been correct in asserting that we will never really know the *whole territory*.

What about technology? We live in an era of technological explosion and there is no reason to believe that it will end anytime soon. The only way technological development might slow down is if our planet can no longer sustain it. Technology is currently being developed by both publicly funded organizations and private companies. It is driven by four factors:

1) Everyday human needs like new agricultural products and foods, medicine, housing, furniture, appliances, electricity, heating, clothing, educational tools, transportation, communications, ma-

chines for various purposes, computers, infrastructure, construction, improved materials, art, entertainment, etc.

2) Scientific needs for new instruments, measuring and recording equipment, simulation, data analysis, data mining, computers and computer programs, space exploration, etc.

3) Demands from the military establishments. It is well known that major technical innovations like radar, nuclear weapons, Arpanet (precursor of the Internet), satellites, robots and drones have been triggered by military demand and funding. Otherwise scarce funds always seem to become available when defense needs are invoked. We seem to be incapable of facing the fact that military developments that look "necessary" at any given time will probably hurt us in the end.

4) The "invisible hand" of the market economy develops many of these products but the profit motive also generates enormous amounts of "stuff" that we don't really need, like thirty different types of cereals and breads, high-sugar soft drinks, junk food, useless clothing and beauty products, new eye-glass fashions, tobacco products, a new cell phone every six months, mindless movies and video games, etc.

In this respect, so-called "virtual reality" (V.R.), engineered to tap into the channels of our *representative reality*, will increasingly invade many aspects of our life experiences. It evolved from relatively passive drawings, paintings, sculpture, and photography to movies and cartoons, and more recently to three-dimensional holography and printing. Immersing people into digital simulations of life-like situations has all kinds of potentials, good and bad. V.R. for training purposes in surgery, engineering and sports may turn out to be very useful. It may also be beneficial for the treatment of certain mental illnesses like PTSD. Unfortunately, V.R. will also be used on a large scale for warfare training. And of course, V.R. has already begun to invade the market of addictive video games and other forms of entertainment that in my mind are of dubious value.

Meanwhile, in connection with the first factor listed above, the world population will continue to grow until at least 2050 and put stress on our way of life and our economies. Today, about half of the world's population of over 7 billion lives in urban centers, and the other half in rural areas. By 2050, with a world population of perhaps 10 billion or more, urban dwellers are projected to make up two-thirds while rural populations will decline to one-third. Even if the world population by then has reached a plateau, the trend towards further urbanization will continue. Making our future cities livable and sustainable will be one of our greatest technical and administrative challenges. Transporting adults from home to work and back, making work available, taking children to school, delivering water, energy, food, and supplies to consumers, will require major new approaches. Will the driverless car work and fulfill expectations?

Automation and robotics for manufacturing as well as retail and office work will greatly change and threaten employment. Some studies predict that in the next twenty years, 50% of current jobs in the U.S. will be replaced by machines. However, nobody seems to know how many new jobs will be created to replace them and how many will disappear altogether. What is likely is that some of the new jobs will require further training and will be well-paid while others will be "lower-skill" jobs paid much less than the ones they replace. One controversial idea being floated by some economists[52] is that under these circumstances, all members of society should receive a universal basic income (U.B.I.) from the state, regardless of whether they have a job or not. Small localized experiments of this type have been tried, for example in Finland, but implementa-

52. See for example Philippe Van Parijs and Yannick Vanderborght, "Basic Income: A Radical Proposal for a Free Society and a Sane Economy," Harvard University Press, 2017.

tion on a national scale is fraught with many social and economic difficulties. It's easy to imagine the controversies that proposing such an idea might generate in the U.S. Congress.

The development of new chips, sensors, ever increasing computing power, gigantic data bases, artificial intelligence, machine learning, and quantum technology will continue to expand our ability to access, use, and manipulate information for both war and peaceful purposes. This said, perhaps instead of spending so much money and effort to develop **artificial** intelligence, we should devote more resources to help our children harness their **natural** intelligence for their self-development, as discussed below.

Education, Our Woes and Psychology

However science and technology develop in the future, one prediction can be made with certainty: education will be more important than ever in the world to ensure decent standards of living and to decrease inequality. This issue raises many questions: What is the minimum number of years that a human being will have to spend in school? What will have to be taught in both developed and developing countries? Who will have to pay for more comprehensive education? What will we require of teachers?

Today, even before acquiring a professional skill in a modern country, it takes a child from birth to the age of 18 to learn the basics of reading, writing, mathematics, and a minimum of the humanities and sciences to function in any society. It is difficult to find exact statistics but it appears that while in the OECD countries about 60% of the adult population has a high school diploma, in China it is closer to 20% and in poorer countries like in Africa, it drops well below 10%, particularly in rural areas. The percentages are even lower if you count those who spent two more years up to the age of ~20 to get a degree from a vocational school, community college, or equivalent. Considerably smaller yet in percentage are those who pursued an undergraduate university education after high school to acquire a degree in engineering, the physical and social sciences, the humanities or arts, or graduate degrees in medicine, law, architecture, business, advanced science, or engineering.

In an ideal world, all education up to the age of 20 should be free and paid with public funding, but this is far from the case, in rich as well as in poor countries. Beyond that age, colleges and universities are becoming much too expensive, worldwide. Their cost should be reduced and they should be at least partially subsidized. In addition, scholarships and combined study-work programs should be made available, unless parents are wealthy enough to afford to pay for their children's education beyond high school.

In addition, beyond the professional training that is needed to earn a livable wage, there is yet another huge deficiency that today's education systems either do not address or address insufficiently. From early childhood on, we humans encounter difficulties that our parents can't always remedy. Even though many of us end up leading relatively satisfying lives, most of us don't get everything we need and want. We may have health problems, feel insecure, dissatisfied, and unable to learn or make positive decisions. Many people remain poor or depressed, and/or suffer from alcohol and drug abuse. We often misbehave with others and discriminate against people different from us. We fight with siblings, parents, friends, colleagues, and strangers. We marry and then divorce. We don't know how to live harmoniously in our communities or how to become good citizens. We may become extremely violent and end up in jail. At some point early on or toward the end of our lives, we may become disabled.

Is this our inevitable destiny? Religious codes were in part invented to help us avoid some of the above pitfalls. Christians are supposed to learn from the Bible, Muslims from the Koran, Jews from the Torah, Gandhi tried to teach non-violence to millions of Hindus, and so on. But as I noted earlier in this book, the fear of God is not enough to control our misbehaviors, and religions are unable to provide all the needed guidance. Psychoanalysts, psychiatrists, and psychotherapists can sometimes help people in distress but it is generally *a posteriori*.

It seems to me that whatever type or level of education we receive, we need to add another important feature to our educational systems, a **preventive component.** Over the last 150 years, famous pioneering educators and their disciples have thought about this problem. American philosopher, psychologist and educator John Dewey (1859-1952) was a champion of education for the development of democracy, participation in civil society, equality for women, and many other social causes. Italian physician Maria Montessori (1870-1952) developed a system based on the observation that young children already exhibit a propensity for self-discipline and that education should build on this trait and encourage them to develop independence and self-reliance. Swiss clinical psychologist Jean Piaget (1896-1980), through keen observations of his own children, studied the various stages of human cognitive development and recommended that education be adapted to the abilities of the learners.

Prevention to avoid some of our woes requires that we go beyond these recommendations, worldwide. Parents are indispensable but they can't do everything. Pre-school should be attended by all children, and pre-school teachers should be trained not just to teach skills but to show children how to become independent as well as cooperative. If children show any signs of dysfunction, these problems should be discussed with their parents and addressed immediately. Psychological help should be provided routinely from kindergarten through high school, and even university. Children and adolescents should be taught how to help others, be compassionate, learn how to negotiate to resolve conflicts, and avoid violence. It is not enough to just teach them sex-education. Adolescents are insufficiently prepared for what they get into when they form bonds with others and fall in love with members of the same or opposite sex. They don't know how to deal with envy and jealousy. They have no early warning of the difficulties involved in marriage and living with another human being, day-in and day-out. They don't get enough help choosing a profession. Most of them are poorly trained to handle money. Except for what they hear from their parents, they rarely know anything about politics and lack the background needed to vote in an informed manner. A huge pedagogical effort lies ahead if one truly wants to better prepare young people for life in addition to a profession. This will require training all kinds of specialists and creating extra financial resources to support this type of curriculum. Even if costly, humanity will save money in the long run. Universities could make an important contribution to formulating such programs by giving a boost to the social sciences. Maybe, one of the best remedies would be to create a compulsory "youth corps" worldwide where all young people would spend a year after high school doing community service somewhere: a universal extension of John F. Kennedy's Peace Corps. Wouldn't this be a tremendous educational and socially-leveling experience?

Finally, how should teaching be done? Technology is already changing many of the tools available to educators with computers, on-line courses and recorded audio and video teaching materials. Google and its competitors are currently making major inroads into schools with Google Docs, Chrome book tablets, the "Classroom" management program, or equivalents. All these new technical aids might reduce the cost of education to some extent and be replicated to help developing countries. However, no technology will ever be able to replace inspiring and passionate educators who can challenge their students and excite their curiosity. Devoted and empathetic teachers and professors have a huge stake in the future of society and must be respected and rewarded accordingly. This is true in Silicon Valley and in the poorest areas of Africa.

Politics, Elections, and Economics

Contemplating the world today, it is not clear where we are going politically. At the end of the Cold War, it appeared that the fairly successful democratic regimes in place in North America, some of Latin America, Western Europe, Japan, India, Australia, and New Zealand would provide a contagious example to Russia, China, South East Asia, Africa, and the Middle East, but this has not happened yet. Autocratic regimes are still prevailing there, and many of the Western-style democracies have been plagued with gridlock, corruption, and back-sliding.

What are the big political problems that history has left at our doorsteps?

Perhaps the most serious one results from inequalities: inequalities of wealth, jobs, resources, social classes, culture, and education, which create cleavages within countries and from country to country. Within countries, these cleavages can result in crime, drug abuse, popular discontent, gridlock, and become the breeding ground for demagogues, rebellions, terrorism, civil wars, and/or religious extremism. Between countries, they create tensions which can lead to trade wars and armed conflict. Resources that could be used for economic betterment are wasted on weapons and military use. Financial inequality could get worse if robots and the "enhanced" humans conjectured by Harari were to relegate large fractions of world populations to menial and poorly paid jobs, or if resources failed to keep up with the population growth, and rationing became necessary!

A second serious problem, even in democracies, is how we elect our leaders. Whether we like it or not, our societies are stratified into classes: the very rich, the middle classes, and the very poor. The first two have their lobbyists, interest groups, and preferred political parties. The very poor, by definition, are weakly represented and can't do much except to march and demonstrate on the street. Leaders and legislators in both presidential and parliamentary systems generally emerge from political parties that ask for their support in political campaigns. To varying degrees, money plays a disproportionate role in these campaigns and distorts their effect and influence. Anglo-Saxon countries generally have two dominant parties that alternate in power. Other countries have a multiplicity of parties and require coalitions to govern. What has recently happened in France with the election of President Macron is another experiment. Other countries yet, like China, Vietnam, and Cuba, have a single party and give their constituents very little choice.

When election time comes, how do people choose to vote, assuming they bother to vote? In principle they vote for the party that seems to represent their values, class, and economic interests, which would make sense. But frequently, they are ill-informed about where their parties stand and what their actual policies will be. They know very little about public policy and economics and get swayed by candidates delivering sound bites on television rather than substantive discussions of issues. As a result, poor and incompetent leaders invoking fear rather than reason, or demagogues with empty promises, can get elected. Numerical electoral margins are often slim and such leaders can get elected with small majorities but do plenty of damage during their terms in office. Are there remedies to these problems in a democratic system? One might be that all potential voters before registering for the first time should be required to take a substantive civics course on public policy and economics. And perhaps, such a course should be required every five or ten years, like the test to renew one's driver's license (not easy to implement!). Another remedy might be for nations to enact rules that, say two or three months before every election, candidates would have to publish a document spelling out in writing what their detailed plans and proposals are, how they would implement them and what they would cost. A similar system already exists in California for ballot measures (so-called state propositions). Even though not everybody reads these voter guides, they are reasonably effective at forcing proponents and adversaries to explain their rationales openly.

As we have seen earlier in the chapter on economics, a third serious problem is that politicians are far too timid when it comes to control booms and recessions that invariably create unemployment. Unemployment, even in the presence of unemployment insurance provided by the state, puts downward pressure on wages, exacerbates income inequality, and sows the seeds of endless social ills. But even in times of booms, the spread in income levels is huge. Raw capitalism pays low-skilled workers wages based on their availability, not on what these workers need for a decent standard of living. Low-paid workers are treated somewhat like not-too-distant descendants of slaves. It seems to me that governments should raise all minimum wage levels to make up for this situation. For example, in the U.S. at the present time, individuals who work 2000 hours/year should earn between $15 and $20/hour, depending on the cost of living where they reside. If employers hire fewer workers at these levels or replace them by robots, the state should hire them for infrastructure or other similar jobs as FDR did as part of the New Deal.

A fourth serious problem has to do with corporations. As I also discussed earlier in detail, corporations, the private engines of the economy, create productive jobs when they feel a demand and see a profit at the end of the line. Unfortunately, together with large banks, they frequently abuse their power to the detriment of society. The "invisible hand" doesn't solve these problems, despite what libertarians would like us to believe. There is no alternative to strong government-imposed protective regulations and accountable bureaucracies to expose corruption.

Barring major disasters, it may be that in another fifty years some of the current inequalities will become less acute on account of total globalization. But this will not be easy. Note that today the total GDP of the world is about $80 trillion which means that the average annual per capita income in the world is barely $11,000. Major improvements will only happen if the world can sustain continued economic growth and if richer countries agree to help poorer ones at the cost of some self-sacrifice. Foreign aid should be increased, not decreased! It is an area where education, social media and communication may eventually level the playing field. But how much time will be needed to overcome religious extremism all over the world? It may happen if communities see an advantage in abandoning sectarianism and violence for economic development and prosperity. The end point should be the universal installation of secular governments, secular education, and freedom of religion, but with complete separation of church and state.

The Future of International Relations

Whether many of these objectives can be realized depends on another major change that I touched upon earlier: the reform of the U.N. As difficult as such a reform may seem at the present time, it is absolutely necessary to the enforcement of international law and the establishment and maintenance of peace in the long run.

The first necessary reform has to do with the Security Council. The five permanent members of the Security Council are too insecure and jealous of their own privileges to relinquish their right to using the veto. But the membership of the Security Council, even with the ten other rotating members, no longer represents the true power relationships and major populations in the world. As a minimum, India, Pakistan, Indonesia, Japan, Germany (or perhaps the entire European Union as a single block), Nigeria, and Brazil should be added as permanent members, as well as another six of the smaller countries that currently have very little voice. With a total membership of perhaps twenty-seven members, it might be possible to retain the veto but allow the full membership of the Security Council to override it by a vote of a super majority of, say, two-thirds of its members.

The second, even harder U.N. reform will be to define thresholds above which a new and more effective Security Council would be allowed to infringe on national sovereignty and intervene in domestic crises. Genocide, use of weapons of mass destruction in a civil war, and large expulsion of refugees ought to be conditions for external intervention. In such cases the enlarged Security Council should have instant access to well-trained *peace-making-and-keeping* armed forces. Its truly international legitimacy would make it superior to the various *coalitions of the willing* cobbled together *ex post facto* in the last few decades for various ad hoc purposes.

A third step that the U.N. should take is to create and train **a respected pool of professional international mediators** capable of intervening (à la George Mitchell in Northern Ireland) in situations where major conflicts and violence break out. Think of conflicts like the Arab-Israeli conflict, the civil war in South Sudan, the civil war in Yemen, or the current tensions in Korea and the South China Sea. They should also intervene in other international issues like nuclear disarmament, global warming, epidemics, and human trafficking and child labor for which institutions are currently too weak to resolve them. Any mediator who succeeds in solving a particular conflict or problem should get a Nobel Peace Prize or equivalent!

Finally, **if I haven't yet sounded utopian enough**, it seems to me that the end-goal of international relations should be to emulate Costa Rica and for all countries to eventually abolish their national armies and turn a remaining fraction of their troops over to a reformed U.N. capable of acting truly collectively.

PART 6:

CONCLUSION: HUMANISM, DEMOCRACY AND HAPPINESS

Let me summarize.

I believe in humanism, not as a religion or just a response to our "feelings." I see it as a set of aspirational goals intended to promote respect for the lives of all human beings on earth through a code of ethics, rights, and responsibilities. Such goals must enable humankind to survive and thrive peacefully worldwide. Humanism must exclude any form of oppression from dictators, ideologies, religions and cults, and yes, Yuval Harari, from computer algorithms. Nor can it and should it assert that we are some kind of master race imposing meaning on the whole universe.

If you want to review a model of international human rights, refer to the Universal Declaration of Human Rights adopted in 1948 by the United Nations. Seventy years later, it is still uplifting! And even if it can never be complete since our conditions evolve, I believe that no system of government will ever be stable unless it observes these rights for all its citizens.

We are not all born with equal characteristics and abilities. We cannot all be Da Vincis, Mozarts, Darwins, Einsteins, Gandhis, Mandelas, or Obamas. We don't all need yachts and private planes, but we all deserve a solid education to understand the world, develop our potential, guide our social behavior, and sustain our livelihood. The current drift towards rising educational and economic inequality is unhealthy and if left uncorrected, will lead to the election of more demagogues, social unrest, and international upheavals. We should acknowledge that not only food, shelter, and jobs are necessary for survival but also healthcare. When 99% of climate scientists tell us that global warming threatens the future of our planet, we should act decisively to try to stem it. If democratically elected politicians refuse to provide universal healthcare or to stem global warming, we should not blame democracy per se but try to correct the ways in which we elect our politicians, the negative influences of corporations, rampant corruption, and money in politics. And as I have indicated in some detail, if we want to put an end to wars, violence, and the current mass displacements of groups and individuals, we must reform the United Nations.

Finally, I want to end with some comments on happiness. The American Declaration of Independence states in its Preamble that we have an unalienable right to the pursuit of Happiness. Happiness is certainly the best state to find oneself in, and endless writers and philosophers have tried to figure out what it is and how it comes about. This reminds me of a little story my father told me years ago.

When my father was growing up at age 17 and studying in Vienna around 1910, he shared a room with a cousin who was a few years older than him. Their room was poorly heated and cold in the winter. As they were lying in their beds at night before falling asleep, they often had philosophical discussions, and one evening, the question of happiness came up. His cousin told my father how he could experience happiness very easily: "Just stick one of your legs out from under the blanket until it gets unpleasantly cold; then just pull it back under the blanket and enjoy the feeling. That is happiness!"

I think the cousin had a point. Happiness can indeed come about as a result of a pleasant change. This would imply that we cannot be happy all the time. Music testifies to this. If for example you listen to a Mozart symphony or piano concerto, you will notice that part of the intellectual and emotional pleasure comes from the way he takes you from a minor (sad) to a major (happy) key, from low to ascending notes, and from a soft pensive melody to an empowering loud crescendo that makes you feel like you are on top of the world. Again, contrast is the secret!

Happiness has a number of approximate synonyms: pleasure, satisfaction, contentment, joy, and fulfillment. Pleasure can be felt when you lie in a hot sun and swallow a cool drink; or when you eat something you really like; or when you are having sex. These are types of happiness that result from our physical nature. They don't last very long, and we need to "pursue" them again and again to be "happy." Buddhist monks think this chase is hopeless. Better to meditate! The root of the word "satisfaction" is "*satis*," which means "the state of no longer being hungry." It is pleasant, particularly if you have been very hungry, but once you have eaten, it doesn't linger around very long. Contentment is the state in which you find yourself after something pleasant has happened, like rejoining a loved one from whom you have been separated. You feel joy when you listen to the last movement of Beethoven's ninth symphony. A dog shows joy when his masters come home. Fulfillment may be a slightly deeper feeling that comes about when a long-time yearning materializes, when one has figured out something after some effort, when one has accomplished something creative, or when one has helped someone else.

All these experiences are temporary. But you know by observing yourself or others that some people at one extreme of the spectrum are always mildly grouchy and unhappy, and at the other extreme, some always radiate a feeling of optimism and happiness (and not because they have just eaten a delicious piece of chocolate or re-encountered a beloved old friend). This difference comes about from their upbringing and their general psychological disposition, both mental and chemical.

The unhappy people find loving difficult, they may themselves feel unloved and insecure, paranoid, incapable of being creative, see tragedy and danger at every corner, and are unable to give. They can feel sad much of the time and never really enjoy the "sunshine." Psychotherapy may help them, hopefully without pills.

The happy people are generally loving, feel secure and positive, are capable of being creative, willing to give and able to express compassion. They may get extremely unhappy and sad at times but they always recover and bounce back.

I don't know where the human condition will be in one thousand years but if all Homo Sapientes became wiser, they would certainly be happier.

ACKNOWLEDGMENTS

Looking back, I want to thank my smart and loving parents Georges M. Loew and Elisabet Loew who took good care of me and helped me get an excellent education from grade school through university. As a result, I was able to find a very interesting and rewarding profession as a scientist. In more recent years, I want to thank my son, George E. Loew, and my brother, E. Sebastian Loew, for their continued support throughout my long effort of writing this book. I also want to thank my step-daughter, Dr. Linda J. Maki, for her help on my chapter on medicine, and Dr. H. Gunther Perdigao for useful discussions on psychoanalysis and Freud. I am indebted to Neil Calder of Okinawa Institute of Science and Technology and San Francisco, and Glenn Roberts (now at LBNL) and Michael Peskin of SLAC National Accelerator Laboratory, who helped me with useful suggestions on a much earlier but related article. I must also admit that without Wikipedia, this book would not have been the same. Finally, I want to thank archivist Jean Deken of SLAC National Accelerator Center for her support and editorial help on my manuscript as well as Gregory Stewart, SLAC graphic designer, for his help in selecting the best images for all my illustrations as well as my book cover.

More generally, I want to acknowledge the enriching conversations and companionship over many years with my wife Gilda, my sister Monique, my daughter Florence, my step-sons Paul and Neil, my daughter-in-law Jane Freston and my son-in-law Troy Surratt, grandchildren, relatives, and long-time friends such as Serge Hurtig from Paris, Carlos Viacava from Buenos Aires and Paris, Sitaram Rao Valluri from Caltech and Bangalore, Guy and Karen Benveniste from Berkeley, California, Richard Pantell from Stanford and San Mateo, California, Anne Knight and Ben Lenail from Palo Alto, California, my "political" companions Emy and

Jim Thurber, Carole Dorshkind, Ashley Evans, and many others. Finally I want to acknowledge the support and friendship of my Ph.D. advisor Karl Spangenberg and my many SLAC mentors and colleagues, including Richard B. Neal, Wolfgang "Pief" Panofsky, Roger H. Miller, Burton Richter, Ewan Paterson, David Burke, Nan Phinney, Jonathan Dorfan, and Juwen Wang.

APPENDIX 1

The Standard Models of Cosmology and Particle Physics

The Standard Model of Cosmology describes our universe as being made of close to 200 billion galaxies consisting of stars, planets, and black holes, which probably started from a singular point with a "Big Bang" 13.8 billion years ago. There is evidence that shortly after the Big Bang the universe underwent an enormous dimensional inflation in a very short time interval, followed ever since by a more gradual expansion. Recent experiments reported in 2013 indicate that all the matter we can see in our universe constitutes less than 5% of its total mass and energy. The remaining 95% is made up of 27% so-called dark matter that we cannot see because it doesn't interact with light (the photons we talked about earlier) but holds our galaxies together through its gravitational attraction, and 68% so-called dark energy which accelerates the expansion of our universe by pulling it apart[53].

At this point we don't yet know what dark matter and dark energy are.

Physicists began to develop the Standard Model of Particle Physics in the second part of the 20th century. It describes and predicts the behavior of the fields and sub-atomic particles that make up the visible world (the above 5%). The sub-atomic particles, six "quarks" and six "leptons" (and their anti-particles made of anti-matter), are visualized as very small points endowed with a collection of properties. They are now in fact considered as small ripples or quanta excited in their respective fields. The particles

53. See European Space Agency, "Planck Reveals an Almost Perfect Universe", March 21, 2013.

interact with each other via four types of forces: the strong nuclear force, the electromagnetic and the weak nuclear forces (now unified under the name of the "electroweak" force), and the gravitational force. Completing this model, two separate large international teams of physicists working at CERN's Large Hadron Collider in Geneva, Switzerland measured the mass of one more particle identified as a "Higgs boson[54]" predicted by as many as six theorists 50 years ago. The theory assumes that this particle is a ripple in yet another field called the "Higgs field" that permeates all of space. The Higgs field, through a complicated mechanism, gives the above particles their respective masses. Two of the principal theorists, Peter Higgs and Francois Englert, received the 2013 Nobel Prize in physics for predicting the existence of this particle and this field.

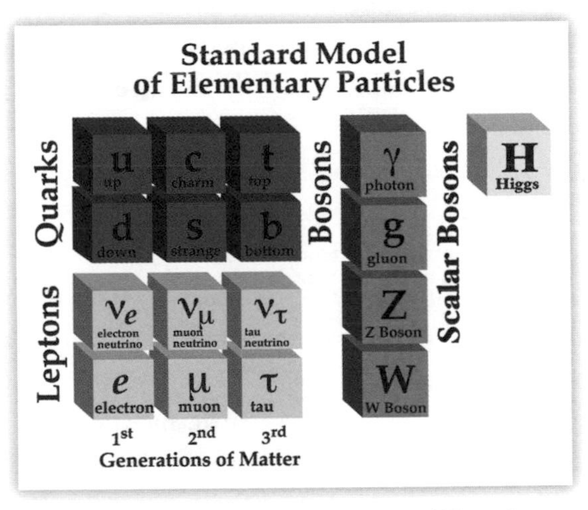

Fig.53 The quarks, leptons, force carrier bosons, and Higgs boson making up the Standard Model of Particle Physics.

---

54. See CERN Courier, Volume 53, Number 4, May 2013 on "Birth of a Higgs Boson," page 21-23. The word "boson" designates a category of particles that obey Bose-Einstein statistics. What this means is that many of them can co-exist in the same physical location or quantum state, as opposed to other particles like quarks and leptons that cannot occupy the same space because they obey Fermi-Dirac statistics.

APPENDIX 2

Time Dilation, Clocks, and Einstein's Principle of Equivalence

In General Relativity (GR), the term "time dilation" can best be understood with respect to the behavior of clocks. Suppose that we start out with two identical clocks, A and B, running at the same rate side-by-side and synchronized at a time t = 0. Suppose now that clock B is taken to another location "B" where for some reason it is caused to speed up with respect to clock A at the original location "A." The time interval between the clicks of clock B actually contracts, which means that it runs faster than clock A (*i.e.*, ahead of clock A). Hence, the time interval between two separate events, the first taking place at point A (*e.g.*, the departure of a light pulse) and the second at point B (*e.g.*, the arrival of this same light pulse), which is the difference between the measured readings of the two clocks, actually increases, or in GR terms, dilates.

One of the elements of Einstein's theory of General Relativity is his Principle of Equivalence. This principle, which this appendix attempts to explain in simple terms, shows that there is an equivalence between the laws of space-time in a stationary reference frame in a gravitational field, like on earth, and those in a reference frame under constant acceleration but in a gravity-free field. The two experiments described below illustrate the "equivalence" of the physics in each case.

Let us start by assuming that we have a rocket of height h = 300 meters. If we place it **horizontally** at rest on the ground of the earth's surface, equip it with two initially synchronized clocks, one at the "bottom" end and the other at the "top" end, and fire a short light pulse from the bottom to the top, the time of flight of the light pulse will be h/c or 1 microsecond (10^{-6} second) as registered by the top clock.

If we erect the same rocket **vertically** on the ground, and do the same experiment by triggering a light pulse from bottom to top, according to GR the change in the gravitational potential away from the earth will distort space-time, and the clock at the top of height h will speed up slightly by a small dimensionless factor gh/c^2. Here g is the acceleration of gravity at sea level (9.81 m/s/s) and c is the speed of light. The top clock speeds up and the time measured by it expands with respect to that measured by the bottom clock by the factor:

$$(1 + gh/c^2)$$

Hence, if the running time of the bottom clock is measured by t, the running time measured by the top clock is:

$$t(1 + gh/c^2)$$

and, in the case of the arrival time of the light pulse from bottom to top, is:

$$(t + h/c)(1 + gh/c^2) = t + h/c + tgh/c^2 \text{ (to first order)}$$

The time interval between the two events is obtained by subtracting t from this expression, so the flight time measured by the top clock is:

$$h/c + tgh/c^2$$

Numerically, if we fire a light pulse from the bottom at $t = 1$ second, it is registered by the top clock at:

$$10^{-6} + 3.3 \times 10^{-14} \text{ second,}$$

i.e., a very small time increment beyond 1 microsecond. We see that the time of travel has dilated a bit!

Furthermore, because of its position, the top clock keeps ticking faster than the base clock. If a subsequent light pulse is fired one second later from the base, the interval registered by the top clock is:

$$1 \text{ microsecond} + 6.6 \times 10^{-14} \text{ second,}$$

and so on. The top clock continues to run away from the bottom clock, and the registered times of arrival continue to grow or "dilate."

The same happens with clocks in **synchronous orbit** satellites, *i.e.*, those that remain stationary above the earth (like the top of the rocket). For the GPS systems that use them, these clocks must be regularly reset (back) by a small amount corresponding to their height above the earth. *The clocks in the satellites are atomic or solid-state clocks that tick at the rate determined by their inherent physical properties in a given gravitational field. They cannot be "slowed down" so they have to be reset to keep the correct time.*

In the second way of looking at the problem, if we now place the same rocket **at rest** in free space (say with respect to some distant star or planet), away from any gravitational field, and fire the same short light pulse from bottom to top, the time of flight will again be h/c = 1 microsecond. *Note once more that the bottom clock ONLY registers the time of departure of the light pulse, and the top clock ONLY registers its time of arrival: neither does both!*

However, if we uniformly accelerate the rocket with acceleration g, *i.e.*, 9.81 m/s/s, again with respect to some distant star or planet (say by firing a retrojet), starting from time t = 0, the rocket, being rigid, doesn't change length, but the top of the rocket moves away from the position of the bottom by an extra distance vt where v is the instantaneous velocity gt of the rocket, and t is the time measured from the instant when the rocket begins to be accelerated from rest (*note that under constant acceleration, the velocity is growing linearly with time*).

If a light pulse is emitted from the base at any time t, it will hit the top after an interval h/c plus the extra time given by:

$$[gt \times h/c] \text{ [extra travel distance] divided by } c.$$

Hence, the time interval between bottom and top clock time readings is again:

$$h/c + tgh/c^2$$

The time dilation is the same as for the static vertical rocket in the earth gravitational field, and by definition the equivalence between the two cases is proven. This is Einstein's Equivalence Principle.

Note that in the second case the two clocks on the rocket are always moving in space with the same acceleration. They are not affected by the acceleration and their rates remain the same because they are in gravity-free space: the clocks in fact do not know anything about the acceleration! It's just that the observer, looking at the time of arrival of the pulse at the top, sees it arriving progressively later than one microsecond with respect to its time of departure and mistakenly thinks that the top clock has sped up!

Of course, the observer also feels an "apparent weight" while standing on the bottom of the rocket. A spring-actuated scale under his/her feet would indicate the same weight that the observer had in the first (static) case on earth because we have assumed in the second case that the rocket has an acceleration g and the observer hasn't gone on a diet in the meantime. Don't get discouraged if you couldn't follow the math! Einstein didn't figure this out overnight.

APPENDIX 3

The History of the Photon or Quantum of Light

About a year before his death in 1955, Albert Einstein lamented in a letter to his friend Michele Besso that although he had thought exhaustively about the photon for fifty years, he still couldn't understand "what it is[55]."

Considering all the revolutionary contributions Einstein made to the concept and properties of the photon, his admission was poignant. Sixty years later, it still raises scientific and philosophical questions that remain interesting to contemplate.

This appendix summarizes how physicists studied the phenomenon of "light" over a period of three hundred years. It shows how they eventually came up with the idea of the ubiquitous "photon" or "quantum of light," and how scientific views on the subject are still evolving. The successive models they developed were based on a series of theories, hypotheses, observations and experiments, and the availability of mathematical tools.

Waves and Particles

We start with Isaac Newton (1643-1727) who believed without proof that light consisted of streams of little particles or "corpuscles." In contrast, his Dutch quasi-contemporary, Christian Huygens (1629-95) and others through the 19th century like Thomas Young (1773-1829), Augustin-Jean Fresnel (1788-1827) and James Clerk Maxwell (1831-79) concluded that light is a wave of

55. "All these fifty years of conscious brooding have brought me no nearer to the answer to the question: What are light quanta? Nowadays every Tom, Dick and Harry thinks he knows it, but he is mistaken. (Albert Einstein, 1954)"

electromagnetic energy propagating in some imaginary medium called "ether," much like water waves on the sea or sound waves in air. Maxwell, through his famous equations, unified electricity and magnetism into the single theory of electromagnetism. Nobody knew exactly what the nature of the ether was and how it could exist in a vacuum, whether it was stationary and the Earth moved with respect to it, or whether the Earth perhaps dragged it along in its motion around the sun. In any case, visible light was understood to occupy just a narrow part of the whole spectrum of electromagnetic waves, to which by definition our eyes are sensitive. All free-space electromagnetic waves, visible light included, travel in vacuum at the same speed (c) and are characterized by a wavelength (λ), the distance between two adjacent crests, as shown below:

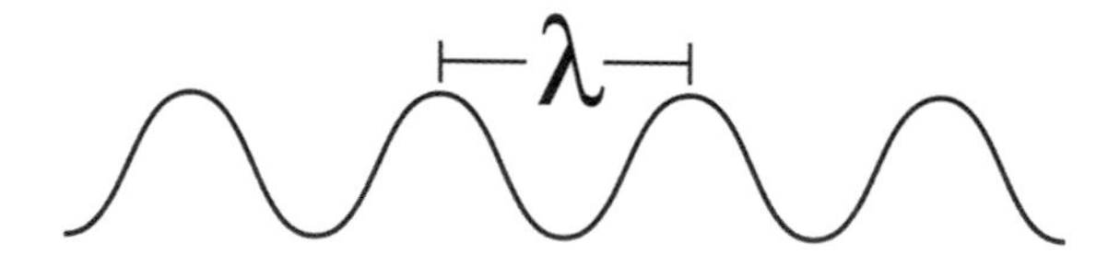

Fig.54 Sinusoidal wave of wavelength lambda.

As we already discussed early on in this book, as the wave travels by at the speed of light, a stationary observer sees the peaks and valleys go by at a given frequency (f) that depends on the distance λ so that the shorter the distance, the higher the frequency. Specifically, the human eye is capable of detecting light with wavelengths ranging from about 0.375 microns (violet) to 0.75 microns (red), where a micron is a millionth of a meter. Since violet light has half the wavelength of red light, it has twice the frequency of red light.

This wave model successfully explained many observations of the behavior of light such as refraction (bending, *e.g.*, by a prism) and diffraction (spreading, *e.g.*, through a hole or a slit), and eventually the propagation of low-frequency radio waves all the way up to very high frequency x-rays and gamma rays. On this basis, it was hard *not* to believe that this model represented "reality" (even though Maxwell apparently stopped short of this belief). Around 1801, Young conducted a seminal experiment that still continues to amaze and puzzle physicists today. The experiment (see figure below) showed what happens when waves from a source of monochromatic light (*i.e.*, of a single frequency) encounter a double slit and split into two separate light beams that then "interfere" with each other. If a screen is placed behind the double slit, this interference produces predictable fringes (peaks and valleys of intensity) on it through coherent addition and subtraction of the two contributing waves. In the figure, the spatial distance between the black circles emanating from the light source and from each of the two slits is equal to the wavelength λ. The fringe pattern is plotted on the right and shows the highest resulting intensity in the middle of the screen where the waves add, as well as minima where they subtract:

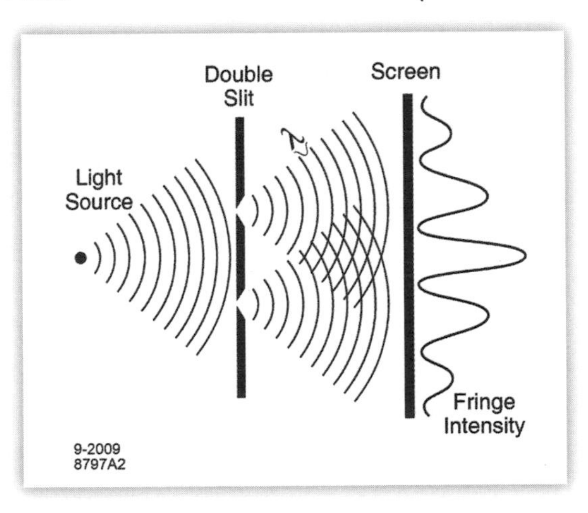

Fig.55 Double Slit Experiment

As compelling as this wave theory was, the story didn't end there. Toward the end of the 19th century, several new observations were made that could not be explained by this theory. One of them was the photoelectric effect whereby light striking a piece of metal can cause it to emit electrons. Another was the intensity and frequency of light emitted by a black body as it is heated up. Indeed, when the temperature of a piece of metal is raised, the intensity of the emitted light increases and its frequency changes, but the old theory couldn't accurately match the experimentally observed spectrum of the glowing light. A third observation was the Michelson-Morley experiment (1887) whose result showed that there was no evidence that the earth moved through a stationary ether or dragged it along. Over the next 25 years, physicists trying to explain these observations spawned new ideas that resulted in one of the greatest revolutions in physics. Widely documented in the scientific and historical literature, this story has been retold recently in the excellent book by A. Douglas Stone (cited earlier in Ref. 15) that gives a new emphasis to Einstein's key influence on the development of quantum theory.

The paradigm shift started in the fall of 1900 when Max Planck (1858-1947) gave two important lectures at successive meetings of the German Physical Society in Berlin. He showed that he had empirically matched the observed radiation spectrum from a black body surface as a function of temperature with a new mathematical model, in which the emitting and absorbing "resonators" (later shown to be molecules) vibrated at discrete (*i.e.*, discontinuous) levels of energy. Albert Einstein (1879-1955), who was just graduating from the Federal Institute of Technology in Zurich, was also puzzled by this problem and over the next few years pushed Planck's model one major step further. By 1905 he proposed that Maxwell's electromagnetic waves themselves consisted of tiny, discrete, massless "packets" or "quanta of energy."[56] He postulated that each packet or quantum is characterized by an energy $E = hf$ where h is Planck's constant and f is its associated frequency, that they are indivisible particle-like objects, and that they travel in vacuum at the speed of light (c), which is constant and independent of the frequency. Some years later, it also turned out that these quanta or photons of a given frequency are characterized by a spin, and that they can be synchronized in phase with each other. The larger the number of photons superimposed in a train (this being possible because they are "bosons," a special property described by Bose-Einstein statistics), the larger the amplitude of the "apparent wave." These ideas inevitably led to the dilemma of the "particle-wave duality," an apparent contradiction that provoked the puzzlement of many physicists who asked themselves: "What is the photon 'really', a particle or a wave, or both or neither?"[57]

Again, using the earlier example, this means that a "violet" photon with twice the frequency of that of a "red" photon has twice the energy of the "red" photon. Einstein argued that an individual photon hitting the surface of a metal can only eject an electron bound in a metallic atomic orbit, and thereby generate a "photoelectric" current in a circuit, if it has enough energy to exceed its binding energy. For the explanation of this effect, he won the 1921 Nobel Prize in Physics. The photoelectric effect is ubiquitous today in hundreds of applications, light meters, cameras, automatic faucets, photovoltaic solar panels, etc.

Together with his work on the photoelectric effect, by 1905 Einstein had also developed his Special Theory of Relativity, followed in 1915 by his General Theory of Relativity, through which

56. It wasn't until much later, in 1926, that chemist Gilbert Lewis called them "photons".

57. Note that the use of Maxwell's equations was in no way negated in their domain of application. For example, the design of the SLAC three-kilometer long copper structure that accelerates the electrons and positrons was entirely done with their help.

he gained worldwide fame. In the Special Theory he postulated that the speed of light has an absolute value that doesn't depend on the reference frame where it is emitted and is constant relative to any observer. This assumption had two fundamental consequences relevant to our discussion here: 1) it eliminated the need for an ether, confirming the validity of the Michelson-Morley experiment and all its subsequent, more refined versions; and 2) it modified our notions of time and space, demoting them from absolute quantities and making them relative to the reference frames in which they are observed. Time intervals and lengths observed in one's reference frame are respectively dilated and contracted when observed in another reference frame in motion with respect to ours. These effects are not illusions: they can be observed through actual physical measurements. As I mentioned earlier in this book, in the process of manipulating formulas and functions in the study of Relativity, Einstein also discovered his famous equation $E_0=mc^2$, which states that the energy of an object **at rest** is equal to its mass multiplied by the square of the speed of light.

Furthermore, as quoted earlier in footnote 15, Einstein played a major role inspiring all the contemporary atomic and quantum pioneers as they came along: Ernest Rutherford (1871-1937), Arnold Sommerfeld (1868-1951), Niels Bohr (1885-1962), Max Born (1882-1970), Erwin Schrödinger (1887-1961), Satyendra Bose (1894-1974), Louis de Broglie (1892-1987), Werner Heisenberg (1901-1976) and Wolfgang Pauli (1900-1958). In 1926 Schrödinger's empirical "wave equation," which successfully predicted the behavior of the hydrogen atom, became the accepted probabilistic "law of the land." Up to a point, Einstein went along with these quantum mechanical views which he had spawned to a considerable extent, but in the end he parted ways with them on philosophical grounds that were discussed earlier in this book.

In the 1940s, another genius, Richard Feynman (1918-1988), arrived on the scene. Along with other members of the new generation that accepted quantum mechanics, Feynman gave up on the idea of absolute determinism. He stated that it was impossible and futile to try to predict the behavior of individual "particles" such as photons. To illustrate this, he revisited Young's double-slit experiment and referred to an amazing experimental observation: even when single photons are emitted one at a time by the source and aimed at the double slit *sequentially*, these photons individually "go" to the back screen and gradually fill out the same fringe pattern. This result seems counterintuitive because one would think that indivisible photons must go through one slit or the other and cannot interfere with "themselves" like Young's waves. And yet, the same interference pattern does form! In all his writings and lectures, Feynman repeated that we would most probably never find "internal wheels or gears" in the photons that predetermine their individual destinations.[58]

Instead, Feynman came up with a modified form of the Schrödinger wave function: "the probability amplitude."[59] He described this concept in detail in his wonderful book for non-scientists, "QED" (for "Quantum Electrodynamics"),[60] and I will try to summarize his explanation here (see Fig. 56) by simplifying it

58. Feynman stated that the photon behaves as if it carries an internal "clock" related to its frequency which "rotates f times" per second.

59. Feynman's concept of probability amplitudes was based on a modified approach to quantum mechanics by which he calculated path integrals along the infinity of possible particle trajectories between two points, a method which he developed in his Ph.D. thesis under John Wheeler at Princeton. Note that the double-slit experiment can also be done with a source of electrons or neutrons and gives the same results as with photons.

60. See Richard P. Feynman, "QED", the Strange Theory of Light and Matter, Princeton University Press, Princeton, New Jersey, 1985. This theory was developed by Sin-Itiro Tomonaga, Julian Schwinger, Richard Feynman and Freeman Dyson between 1946 and 1950.

slightly. It is based on the view that when a particle like the photon goes from point (1) to point (2), it doesn't necessarily move along a straight line but has certain *probability amplitudes* of following not a single trajectory but an infinite number of different broken or "kinky" trajectories to get from (1) to (2). It is as if all these trajectories are "explored" by the photon, as if it had multiple choices![61] Referring to the figure, each possible trajectory, of which I am only showing two, can be represented by a little arrow or "vector" of fixed length, which rotates by 360 degrees at the photon's frequency of emission, like the little stopwatches shown on the left. When you add these vectors together head-to-tail,[62] you get the final probability amplitude for the photon to go from the source to L and then to M, represented here by vector A, or by the probability amplitude for the photon to go from the source to R and then to M, represented by vector B. The paths LM and RM are equal because M is chosen to be in the middle with respect to the two slits, and the two little stopwatches start from the source with the same angles, also called phase angles.

Feynman's rule says that to predict the outcome of the experiment, you must now add the two equal vectors A and B that line up in the same direction, *i.e.*, with the same phase angle as shown. Say each vector has a length of 0.1: the sum of the two is 0.2. Following a rule originally suggested by Max Born, you then take the square of this number, 0.04, and this gives you the probability of finding the photon at M. Similarly, if you go through the same reasoning for the probability amplitudes ending at N, the two vec-

61. See reference 3 above, Stephen Hawking and Leonard Mlodinow, "The Grand Design," Bantam Books, 2010, page 80, where these multiple choices are designated as "alternative histories." When added as explained here, they yield what is called "the sum over histories."

62. If I had used an infinite number of trajectories, the argument and the diagram would be more complicated but the result would be the same.

tors subtract to zero because the paths LN and RN differ by half a wavelength as a result of the distance between the two slits L and R: this means that the phases of the two vectors are 180 degrees apart at the rotation frequency of their "clocks," which makes them cancel. When squared, this results in a "null" or zero probability for the photon to end at N.

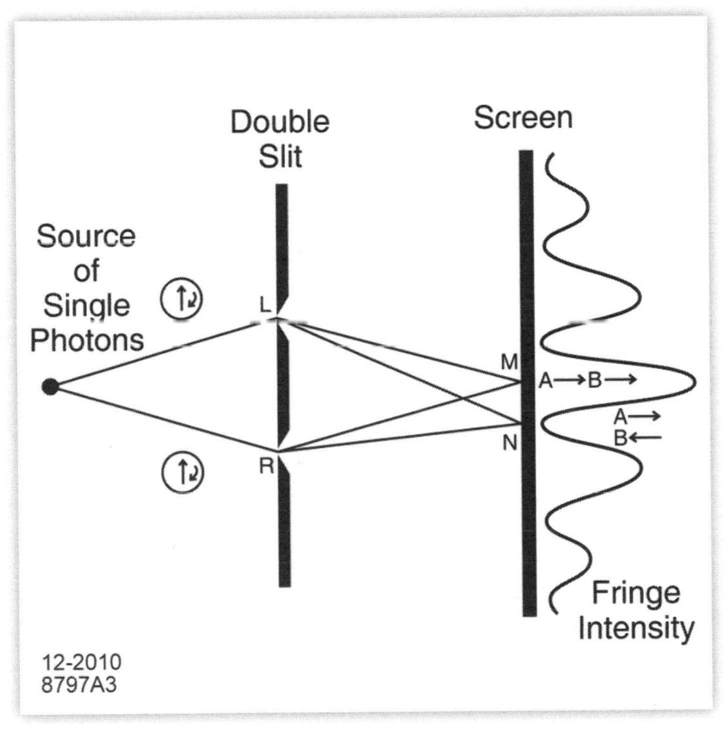

Fig.56 Interference pattern of two or more photons.

Feynman said that by adopting this "recipe," you didn't have to worry about the particle-wave duality dilemma anymore. (It seems, however, that by linking the rotation of the probability amplitude vector of the photon to its frequency f, he had not abandoned the vestige of the wave picture).

With the amazing success of this quantum mechanical approach which is now applied to predict all nuclear reactions and particle decays, Feynman preached that we should accept Mother Nature as strangely as she behaves, limit ourselves to making observations, and be satisfied with models that *predict* these observations, even if they don't give a mechanistic picture or tangible *explanation* of the process (Note the very important distinction here between prediction and explanation of a result). This approach has stood the test of time so far, and it is where physics stands today in this respect. It may not be the end of the story: "qui vivra verra!"

The photon is ubiquitous in our world. It mediates all our communication systems in air, cables, fiber optics, and via satellites; it lights up our environment at night and illuminates our TV and computer screens; it is involved in photo electricity and radar; it accelerates charged particles in accelerators and cooks our food in ovens at microwave frequencies; it is involved in the excitation and absorption of energy of electrons in atoms; it is the intermediate or "exchange" agent that explains the forces between electrons and positrons; it is emitted as "synchrotron radiation" in machines in which charged particles are accelerated, or as "bremsstrahlung" when they are decelerated in targets; it makes up the "monochromatic" (at a single frequency) and "coherent" (at a single phase) beams of lasers; it is used as x-rays in medicine, as a probe in material science, biology, and chemistry; it is central to the life of plants in the process of photosynthesis, and produces global warming through infrared light trapped in the atmosphere. It enables our observations in astrophysics at visible, x-ray and gamma ray (super high energy) frequencies. Finally, it provides us with insight into the cosmology of our early universe through the detection and study of the low frequency cosmic microwave background (CMB).

The CMB is today's relic of the high frequency photons emitted roughly 400,000 years after the Big Bang when fully ionized matter in the form of protons and electrons combined into atoms and became transparent, releasing the trapped radiation and letting it "cool down" (stretching its wavelength) as the universe expanded.

Today, for the record, the speed of light or photons in vacuum, commonly denoted by the letter **c**, is **299 792 458 meters per second,** where the meter is defined from this constant, and the second is given by the international standard for time. This standard is derived from the transition between two hyperfine levels of the ground state of the Cesium 133 atom unperturbed by black body radiation.

Unlike the speed of sound in air, contemporary physicists don't ask why the speed of light in vacuum has the value it has. They accept it as a constant of nature. I wonder why, given all the other weird properties that have been invented for the vacuum in recent years.

CREDITS

Fig.1 Public Domain

Fig.2 Public Domain

Fig.3 Public Domain

Fig.4 Public Domain

Fig.5 Credit: Christopher Michel creator QS:P170,Q5112871 (https://commons.wikimedia.org/wiki/File:Orcas_&_humpbacks_(3730257925).jpg), "Orcas & humpbacks (3730257925)," https://creativecommons.org/licenses/by/2.0/legalcode

Fig.6 Credit: Thomas Lersch (https://commons.wikimedia.org/wiki/File:Schimpanse_Zoo_Leipzig.jpg), "Schimpanse Zoo Leipzig," https://creativecommons.org/licenses/by-sa/3.0/legalcode

Fig.7 Credit: Gregory Stewart/SLAC National Accelerator Laboratory

Fig.8 Public Domain

Fig.9 Public Domain

Fig.10 Public Domain

Fig.11a Credit: Author Steven Lehar

Fig.11b Credit: Author Steven Lehar

Fig.12 Credit: Dhirad (https://commons.wikimedia.org/wiki/File:Taj_Mahal_in_March_2004.jpg), "Taj Mahal in March 2004," https://creativecommons.org/licenses/by-sa/3.0/legalcode

Fig.13 Public Domain

Fig.14 Credit: Salvador Dalí. (1904-1989). *The Persistence of Memory*. 1931.

Fig.15 Public Domain

Fig.16 Public Domain

Fig.17 Public Domain

Fig.18 Public Domain

Fig.19 Public Domain

Fig.20 Credit: Caltech Archives

Fig.21 Credit: R. Hurt/Caltech-JPL

Fig.22 Credit: Caltech/MIT/LIGO Lab

Fig.23 Credit: Sandbh (https://commons.wikimedia.org/wiki/File:18_column_periodic_table,_with_Lu_and_Lr_in_group_3.png), https://creativecommons.org/licenses/by-sa/4.0/legalcode

Fig.24 Public Domain

Fig.25 Royalty-free

Fig.26 Credit: Jewish Chronicle Archive/Heritage Images

Fig.27 Credit: MagentaGreen (https://commons.wikimedia.org/wiki/File:Experimental_setup_of_Photo_51.svg), "Experimental setup of Photo 51," https://creativecommons.org/licenses/by-sa/2.0/legalcode

Fig.28 Public Domain

Fig.29 Public Domain

Fig.30 Public Domain

Fig.31 Credit : Greenhouse Gas by Sector.png: Robert A. Rohde derivative work: Setreset (talk) (https://commons.wikimedia.org/wiki/File:Greenhouse_gas_by_sector_2000.svg), "Greenhouse gas by sector 2000," https://creativecommons.org/licenses/by-sa/3.0/legalcode

Fig.32 Credit: Copyright 2002 Wadsworth Group, a division of Thomson Learning, Inc.

Fig.33a Public Domain

Fig.33b Credit: Robert Huffstutter (https://commons.wikimedia.org/wiki/File:FREUD'S_SOFA.jpeg), "FREUD'S SOFA," https://creativecommons.org/licenses/by/2.0/legalcode

Fig.34 Credit: Gregory Stewart

Fig.35 Credit: Gregory Stewart

Fig.36 Public Domain

Fig.37 Public Domain

Fig.38 Public Domain

Fig.39 Public Domain

Fig.40 Public Domain

Fig.41 Public Domain

Fig.42 Public Domain

Fig.43 Public Domain

Fig.44 Public Domain

Fig.45 Public Domain

Fig.46 Credit: Bibi Saint-Pol (https://commons.wikimedia.org/wiki/File:Map_Greco-Persian_Wars-en.svg), "Map Greco-Persian Wars-en," https://creativecommons.org/licenses/ by-sa/3.0/legalcode

Fig.47 Public Domain

Fig.48 Public Domain

Fig.49 Public Domain

Fig.50 Public Domain

Fig.51 Credit: National Geographic (https://commons.wikimedia.org/wiki/File:02-arkhipov-young.ngsversion.1495227880056.adapt.1900.1.jpg), https://creativecommons.org/licenses/by-sa/4.0/legalcode

Fig.52 Credit: Arms Control Association (https://www.armscontrol.org/factsheets/Nuclearweaponswhohaswhat)

Fig.53 Credit: SLAC National Accelerator Laboratory

Fig.54 Credit: Gregory Stewart/SLAC National Accelerator Laboratory

Fig.55 Credit: Gregory Stewart/SLAC National Accelerator Laboratory

Fig.56 Credit: Gregory Stewart/SLAC National Accelerator Laboratory

INDEX

Note: Page references with an *f* are figures.

E

F

I

J

N

O

P

Q

R

S

T

U

V

W

X

Y

Z